Public Health and Menstrual Hygiene Practices in India

This book examines the equity issues regarding practices of menstrual hygiene and affordability of menstrual products by the lower socioeconomic class in India.

By discussing a novel and detailed methodology of estimating the cost of menstrual hygiene, the authors identify various components of direct and indirect costs of menstrual hygiene. The research makes use of a mixed-methods approach to identify key cost drivers to estimate the overall costs across adolescent/young and older women belonging to diverse socioeconomic status. It also discusses the relevant socio-cultural issues associated with menstruation and menstrual cycles, as beliefs and taboos associated with menstrual cycles are the second most deciding factors influencing the choice of menstrual products after affordability.

This book will be of interest to researchers in the field of Women's and Reproductive Health, Public Health in India, Health and Wellbeing and South Asian Studies.

Smruti Bulsari is a Senior Research Officer at the University of Essex, where she works on National Institute of Health and Care Applied Research Collaboration (NIHR - ARC), and Alzheimer's Society-funded research on dementia. She has worked on research projects funded by ESRC, British Academy, ICSSR and Government of Gujarat. She has a PhD in Economics and has experience working with large datasets.

Kiran Pandya is Vice Chancellor of Sarvajanik University. He was also appointed as the Member, National Statistical Commission (NSC), Ministry of Statistics and Programme Implementation (MoSPI). He was a Professor and Head in the Department of Human Resource Development, Veer Narmad South Gujarat University, Surat (Gujarat, India). He obtained the Doctor of Philosophy degree in Economics by the University of Sussex (UK), for which, he was awarded the Academic Staff Scholarship by the Commonwealth Commission in the UK.

Anil Gumber is a Visiting Professor in health economics and medical statistics at the Research Centre for Healthcare & Communities, Coventry University (UK). He also holds an Emeritus Senior Fellow position at the Faculty of Health and Wellbeing, Sheffield Hallam University. He holds a PhD in Economics and was a postdoctoral research scholar at the Harvard School of Public Health, Boston.

Routledge Contemporary South Asia Series

Foreign Aid and Bangladesh
Donor Relations and Realpolitik
Mohammad Mizanur Rahman

Empowering Marginalised Women in Remote Indian Villages
An Impact Study
Udoy Saikia, Jim Chalmers, Dency Michael, and Janice Orrell

Marginalised Groups in India
Historiography, Politics, and Policies
Edited by Kunal Debnath and Souvik Chatterjee

Creative Economies of Culture in South Asia
Craftspeople and Performers
Edited by Anna Morcom and Neelam Raina

Everyday State and Development in Northeast India
Biswaranjan Tripura

Political and Social Change in Pakistan's Tribal Areas
Understanding Pashtun Youth
Naveed Ahmad Shinwari

Identity, Dispossession and Resilience of the Subaltern
A Study of Marginalized Communities in Kashmir
Khalid Wasim Hassan, Deepanshu Mohan, Ishfaq Ahmad Wani, and Najam Us Saqib

Public Health and Menstrual Hygiene Practices in India
Practices, Costs and Equity Issues
Smruti Bulsari, Kiran Pandya and Anil Gumber

Public Health and Menstrual Hygiene Practices in India

Practices, Costs, and Equity Issues

Smruti Bulsari, Kiran Pandya and Anil Gumber

Routledge
Taylor & Francis Group

LONDON AND NEW YORK

First published 2025
by Routledge
4 Park Square, Milton Park, Abingdon, Oxon OX14 4RN

and by Routledge
605 Third Avenue, New York, NY 10158

Routledge is an imprint of the Taylor & Francis Group, an informa business

British Library Cataloguing-in-Publication Data
A catalogue record for this book is available from the British Library

Library of Congress Cataloging-in-Publication Data
Names: Bulsari, Smruti, author. | Pandya, Kiran, author. | Gumber, Anil, author.
Title: Public health and menstrual hygiene practices in India : practices, costs and equity issues / Smruti Bulsari, Kiran Pandya and Anil Gumber.
Description: Abingdon, Oxon ; New York, NY : Routledge, 2025. |
Series: Routledge contemporary South Asia series | Includes bibliographical references and index.
Identifiers: LCCN 2024062072 (print) | LCCN 2024062073 (ebook) | ISBN 9781032754338 (hbk) | ISBN 9781032754345 (pbk) | ISBN 9781003473961 (ebk)
Subjects: LCSH: Menstruation--Social aspects--India. | Menstruation--Economic aspects--India. | Women--Health and hygiene--India. | Public health--India.
Classification: LCC QP263 .B943 2025 (print) | LCC QP263 (ebook) | DDC 612.6/620954--dc23/eng/20250311
LC record available at https://lccn.loc.gov/2024062072
LC ebook record available at https://lccn.loc.gov/2024062073

ISBN: 978-1-032-75433-8 (hbk)
ISBN: 978-1-032-75434-5 (pbk)
ISBN: 978-1-003-47396-1 (ebk)

DOI: 10.4324/9781003473961

Typeset in Times New Roman
by SPi Technologies India Pvt Ltd (Straive)

Dedicated to
All women (whether or not they are menstruators)
and
All menstruators (whether or not they are women)

Contents

Figures

Tables

Foreword

In recent years, the conversation surrounding menstrual health and well-being has gained significant momentum, yet millions of individuals in the UK – and indeed across the globe – still face barriers that prevent them from accessing the necessary products, education, and support. The ramifications of this inequity extend beyond the individual, leading to increased absenteeism in schools and workplaces, deepening feelings of shame, and resulting in various psychological impacts. Addressing these issues is not just a matter of individual health; it is a societal imperative that benefits us all.

The research compiled in this book offers an invaluable exploration of how people access information regarding menstrual health and hygiene. It sheds light on the physical and mental toll that inadequate support can inflict, not only emphasising the importance of this subject but also urging a shift in societal attitudes toward menstruation. One stark finding reveals that 1 in 10 menstruators have been discouraged from discussing their periods in public, a phenomenon often exacerbated by cultural stigmas.

The book's authors have approached this critical topic with a meticulous and compassionate lens, providing insights that have the potential to inform and influence local policy changes. By examining the direct and indirect financial costs associated with menstruation, this research quantifies the economic burden on individuals, offering a persuasive argument for the need for systemic change.

As a CEO dedicated to championing user voice and lived experience, it is heartening to witness the depth of information gathered throughout this exploration. Although the research is based in Gujarat, India, the principles and policy recommendations resonate widely, reflecting a shared reality that transcends geographical boundaries.

The discussions around the types of sanitary products used reveal significant societal shifts, highlighting how financial constraints, educational opportunities, and environmental concerns shape decision-making processes. Notably, the 2024 initiatives in the UK – such as the school period product scheme and the removal of VAT on period underwear – offer a robust framework for supporting young people in accessing essential products, while also contributing positively to the environment.

Yet, despite progress, cultural taboos and misunderstandings continue to prevail. Our work supporting men, including single fathers, has demonstrated that many struggle with discussions around menstruation, creating a cycle of silence that fosters shame and guilt. It is crucial to recognise that much of the menstrual knowledge is passed down through maternal figures. However, when misinformation is present, it perpetuates a legacy of misunderstanding.

A central theme emerging from this well-referenced research is the stark reality of how menstruation – an experience shared by the majority of women, transgender men, nonbinary, and genderqueer individuals – contributes to widening inequalities. The book concludes with examples of successful practices implemented by NGOs and pilot programs in India, emphasising the importance of learning from these initiatives to enrich the dialogue around subsidised sanitary products and safe hygiene education.

By engaging in honest conversations about menstruation, we can collectively dismantle the taboos and misinformation that persist, paving the way for a more equitable future. Let this book serve as a catalyst for change, encouraging all of us to advocate for better understanding and access to menstrual health resources. Together, we can work toward bridging the inequalities gap and creating an inclusive society for everyone.

Samantha Glover
CEO, Healthwatch Essex
Did Ed, BSc (hons), MSc, UKPHR

Preface

Almost fifteen years ago, Kiran Pandya and I began our research for informing district-level plans for two diametrically opposite, yet geographically adjacent, districts of Gujarat (India). Several dimensions of human development were studied over a period of close to 7 years. One of those dimensions was "health". We had a team that did survey work to complement the data already available with the local authorities. While we researched a plethora of health issues, we soon realised that menstrual health and hygiene need attention as much as other health issues. World Bank statistics show that on any given day, more than 300 million girls/women menstruate across the world. On the other hand, at least 500 million women worldwide lack access to manage their periods. In India, 12 per cent of 355 million menstruators cannot afford menstrual products, and more than 30 per cent of girls/women do not use hygienic methods for managing their periods.

Addressing period poverty, defined as lack of affordability of period products, is absolutely essential for ensuring good menstrual health. It also has a positive externality in terms of increasing the Gross Domestic Product (GDP) of India by 2.7 per cent. Menstrual health investment index is a recently developed tool to assess the hygiene practices and menstrual health of young adults. India's menstrual health investment index is at 1.2 per cent of India's GDP, which requires increasing to address period poverty.

The scope of our then-ongoing study constrained our exploration into the issues of menstrual health beyond a point. Nevertheless, we did manage to set up focus group discussions with girls and women of different socioeconomic strata to build a strong rationale for exploring the research on menstrual health in depth. In 2017, we submitted an application to the Indian Council of Social Science Research (ICSSR), New Delhi, to fund this study. It took nearly a year, and then we were invited to ICSSR, New Delhi, for a presentation to defend our research proposal. Since I was out of the country during those days, Kiran Pandya and Minasree Saikia (my then colleague) defended the proposal; it was all face-to-face until the pandemic struck. We are thankful to ICSSR for generously funding our research.

Kiran did all the groundwork, formed the team, and contacted some doctors and non-government organisations (NGOs) for the patients and public involvement (PPI) work. Malabika Mahajan, the then computer lab coordinator, along with our librarian, Dharma Naik, helped in generating the bibliography to be explored for developing the questionnaire and interview schedules for our research.

Unfortunately, for various reasons, this project was a non-starter until January 2020. The project team had to face a lot of adverse circumstances. I was not able to return to India at least a couple of months since we got the funding. The research assistant was recruited, but she left the job within a couple of weeks.

We finally began the PPI work in January 2020. Many girls and women in the menstruating age group contributed towards the development of our questionnaire. We learnt a lot about the pre-menstrual syndrome (PMS) and ailments experienced by women during the menstrual cycles. The questionnaire was revised around 3–4 times alongside the PPI team of doctors and girls/women and translated to Gujarati, the vernacular of Gujarat State, before it was presented to the reviewers appointed by the ICSSR. We are thankful to Bhavna Patel for translating the questionnaires to Gujarati. The reviewers appointed by the ICSSR were Madhusree Shekhar, Professor and Chairperson, Office for International Affairs, Tata Institute of Social Sciences (TISS), Mumbai, and Niti Mehta, Professor and Director, Sardar Patel Institute of Economics & Social Research (SPIESR), Ahmedabad. They reviewed the methodology of our study and the survey instruments. They appreciated our initiative and gave some valuable suggestions, which we incorporated in our questionnaire. We then pilot tested the questionnaire and made further edits based on the issues faced in administering the questionnaire. Despite being a late starter, we were ready with the pilot-tested questionnaire by February 2020. But this time, the entire world was faced with the worst pandemic of their lifetimes. The entire world was under lockdown, and so was our study. The lockdown situation continued, and we decided to begin the survey work, with appropriate changes to the methodology and sampling, and employing online questionnaires. Everyone was gradually adapting to the "new normal", and so did we. We thank Rashmita Nayak, who developed the online versions of both, Gujarati and English, questionnaires using Google Forms. Minasree, Kiran, and I pilot tested them again.

The journey of our research would not have been possible without the support of people from different walks of life. We would like to thank all those who helped us during the journey of this research work and even after. Our PPI team comprised of teaching assistants: Firuzi Bhathena and Yesha Joshi, plus some girl students of the Department of Human Resource Development, Veer Narmad South Gujarat University, Surat.

We thank the anonymous NGOs who gave valuable details about their interventions in improving the menstrual health of people belonging to the lower socioeconomic strata, and also the anonymous participants of in-depth interviews who shared their most intimate lived experiences to give very valuable insights into the menstrual health issues, including menstrual dysfunction.

We also acknowledge the support of the joint coordinators of the project, Jayesh Desai and Gaurang Rami, who helped with screening the articles extracted by Malabika Mahajan and Dharma Naik and for dissemination of outputs. The field investigators did an excellent job of data collection through telephonic and later on, face-to-face surveys, after the situation improved. Our field investigators are: Arti Parmar, Akshata Jain, Bhavika Makwana, Bhumika Makwana, Bhavna Patel, Dharma Naik, Jagruti Parmar, Kajal Singh, Kinjal Valand, Sapna Singh Santraj, Shruti Arthaniya, Shital Sharma, Reema Parekh, and Vijeta Patel. Rushang Bhandari meticulously undertook the herculean task of data entry and data cleaning, which was supported and cross-verified by Minasree. Divyani Patel transcribed the recorded interviews.

We also lost some of our important PPI members on the journey. We received immense support in developing the questionnaire from Late Dr Sudevi Hazari, a renowned gynaecologist of Surat city, and Late Dr Minaxi Desai, an Ayurvedic doctor specialised in women's reproductive and intimate health, also from Surat city. May their souls rest in peace.

We are extremely grateful to our publishers, Taylor & Francis, for considering our work for publication and their support throughout the publication journey. We would also like to thank the anonymous reviewers for their comments on the book proposal and sample chapters. It did help a lot in improving the draft and reorganising the book chapters to make it reader-friendly.

Finally, we would like to thank our respective family members for their valuable support.

Smruti Bulsari
December 2024

Acknowledgement

Kiran Pandya and Smruti Bulsari would like to thank the Indian Council of Social Science Research (ICSSR), New Delhi for funding this research. Reference: **File No. G-59/2017-18/ICSSR/RPS**

1 Introduction

Menstruation is referred to as a "normal biological process" in the literature on menstrual health. Winkler (2020) explains that cultural, political, social, geographical, religious, and health-related factors shape the rights of menstruators[1]. In other words, awareness, beliefs, and taboos, around menstruation and accessibility to menstrual products influence the menstrual management practices. Menstrual management practices, menstrual hygiene, and menstrual health go hand in hand. Therefore, it becomes important to understand the factors influencing the menstrual management practices of menstruators. Recently, WHO has directed that menstruation be given the status of health issue, rather than just hygiene issue. They have further directed providing education and information to menstruators about menstruation, ensuring that the activities surrounding menstrual education and information are included in the relevant work plans and budgets, and their performance monitored (WHO, 2022).

Our study attempts to understand menstruation management practices in India and how sociocultural factors like awareness, beliefs, and taboos around menstruation shape these practices. We also hypothesise and eventually observe that economic factors like the cost of menstrual management products and accessibility pose major constraints to hygienic menstrual management practices, especially to the menstruators belonging to lower socioeconomic strata. Data from the 5th National Family Health Survey (NFHS) of India shows that girls in the highest wealth index[2] are twice as likely to adopt hygienic methods as compared to those in the lowest. This difference was much higher (almost 3.5 times) in the 4th round of NHFS (MoHFW and IIPS., 2020); NFHS-4 was conducted during 2015–2016 and NFHS-5 during 2019–2020.

At this juncture, we would like our readers to know that the data collection phase of our study coincided with the onset of COVID-19 pandemic and subsequent lockdown. Therefore, our study captures the additional constraint to accessibility posed by the lockdown and disrupted supply chain. Supply of sanitary napkins, tampons, and menstrual cups could have been affected during the lockdown because they did not

DOI: 10.4324/9781003473961-1

form a part of the essential commodities when this study was undertaken. Later, they were proposed to be included in the list of essential commodities in The Essential Commodities (Amendment) Bill (2022). Nevertheless, the Government of India had abolished the Goods and Services Tax (GST) on sanitary napkins in 2018 (Press Information Bureau, 2018), thus making it more affordable since then.

This study was undertaken to understand the demand-side constraints – economic as well as sociocultural – that hinder hygienic menstrual management practices. The results of this study would provide insights to policymakers and charity organisations for developing and implementing menstrual health policy for the citizens. Other developing countries with similar sociocultural and economic backgrounds can also derive similar insights and take necessary action for improving the quality of life of menstruators. It might be inspirational for the policymakers to learn from Scotland, where it has been made available as free period products by implementing the Period Products (Free Provision) (Scotland) Act (2021).

This chapter is organised into five sections: in Section 1.1, we present some statistics to explain the situation of menstruation management in India using the NFHS data. While, we have already given a background on why this study is important, a detailed rationale for undertaking this study is described further in Section 1.2. The methodology adopted to undertake research and adaptation of data collection method to suit the lockdown conditions is presented in Section 1.3. In Section 1.4, we describe the socioeconomic and demographic characteristics of the respondents of this study, and finally, a brief overview of the rest of the chapters of this book is given in Section 1.5.

1.1 Menstrual Management Scenario in India

The results of NFHS 4 and 5 show the usage pattern of the absorbent, used for soaking menstrual blood, by the post-menarche adolescent girls in the age group of 15–19 years[3] (UNFPA, 2022). Aggregate data shows that there is an increase of 20 percentage points in the number of girls using any hygienic method to soak menstrual blood, in the 15–19 age group, from 2015–2016 to 2019–2020 (from NFHS-4 to NFHS-5). Hygienic methods to soak menstrual blood include locally prepared napkins, sanitary napkins, tampons, and menstrual cups (MoHFW and IIPS., 2020). The data also shows that the percentage of girls using cloth as an absorbent has decreased from 62.5 per cent to 49.3 per cent.

Before proceeding further, we would like to clarify that menstruation management is a broader term, encompassing methods of soaking menstrual blood, cleaning private parts (vagina and vulva, to be specific), and washing hands. Therefore, the term "menstrual management" is NOT used interchangeably with "method of soaking menstrual blood". In this section, we specifically discuss the methods of soaking menstrual blood.

In urban areas, 90 per cent of these post-menarche adolescent girls use hygienic methods of absorbing menstrual blood as compared to 73 per cent in rural areas in 2019–2020. This is an increase from 79 per cent and 40 per cent in urban and rural areas respectively, from 2015 to 2016. This indicates that rural girls are quickly adopting hygienic methods to soak menstrual blood. The data of both NFHS 4 and 5 show that the percentage of girls adopting hygienic methods to soak menstrual blood is directly associated with their education level, and girls with higher (than secondary-level school) education are twice as likely to use hygienic methods as compared to those below secondary level.

A higher proportion of unmarried girls (79 per cent) are found to be adopting hygienic methods as compared to married girls (72 per cent) in the NFHS-5 survey. In NFHS-4, these percentages for unmarried and married girls were 60 per cent and 48 per cent respectively. This implies that the difference in percentage points of unmarried and married girls using hygienic methods has reduced from 22 per cent in 2015–2016 to 7 per cent in 2019–2020.

The distribution of hygienic methods of soaking menstrual blood shows that girls belonging to religions "other[4]" than Hindu, Muslim, or Christian is the highest (89 per cent), as per NFHS-5 survey. This is only 2 percentage points higher than Christian girls, 87 per cent of who use hygienic methods. Hindu and Muslim girls using hygienic methods to soak menstrual blood are much lower (78 and 75 per cent respectively) as compared to girls from "other" religions. Also, the increase in percentage points of girls adopting hygienic methods from NFHS-4 to that in NFHS-5 is also highest in "other" religions as compared with that in Hindu, Muslim, and Christian religions. This increase is 15 percentage points, for girls belonging to "other" religions, from 74 per cent in NFHS-4. Hindu, Muslim, and Christian girls adopting hygienic practices in NFHS-4 had been 58 per cent, 54 per cent and 79 per cent respectively.

The highest percentage of girls using hygienic methods to soak menstrual blood is 86 per cent in NFHS-5, and they belong to the general category. This is an increase from 69 per cent in NFHS-4. This percentage for schedule caste/schedule tribe (SC/ST) and other backward class (OBC) girls is 75 and 78 respectively, which is an increase from 52 and 57 per cent respectively, from NFHS-4.

Among the hygienic methods used to soak menstrual blood, the most popular product happens to be the disposable sanitary napkins (64.5 per cent).

Even until NFHS-5, 0.2 per cent girls do not use any absorbent to soak menstrual blood. In the patient and public involvement (PPI) activity, undertaken for eliciting the menstrual hygiene and health issues, we learnt that girls are taught at an early age to restrain the blood flow and release it at regular intervals, so that they can manage without any absorbent. While this percentage has reduced from 0.4 (NFHS-4), the absolute numbers work out to 48,164 for NFHS-4 and 24,361 for NFHS-5. These are worrisome figures. These data are for adolescent girls only, who are "new menstruators" (15–19 years age). It may be noted that detailed data and explanation on menstruation practices for the girls in the age group of 15–24 years can be found in MoHFW and IIPS. (2020).

It may be noted that with every passing year of menstruation experience, the menstruators are more likely to settle down with a product/absorbent and might be less likely to experiment with newer methods for soaking menstrual blood. Most studies show that girls first came to know about menstruation from their mothers, and also received guidance about menstruation management from their mothers (Chaghlana et al., 2019; Patel and Patel, 2016; Santra, 2017). Even our study shows that mothers are the first source of information about menstruation and its management. Thus, an increase in percentage of new menstruators using hygienic methods as absorbents is also a reflection of a shift in mothers' attitudes towards menstrual hygiene practices.

Sengupta et al. (2020) report consolidated percentages for women in the age group of 14–49 years, by state, to highlight regional variations in hygienic methods of soaking menstrual blood. Figure 1.1 gives an

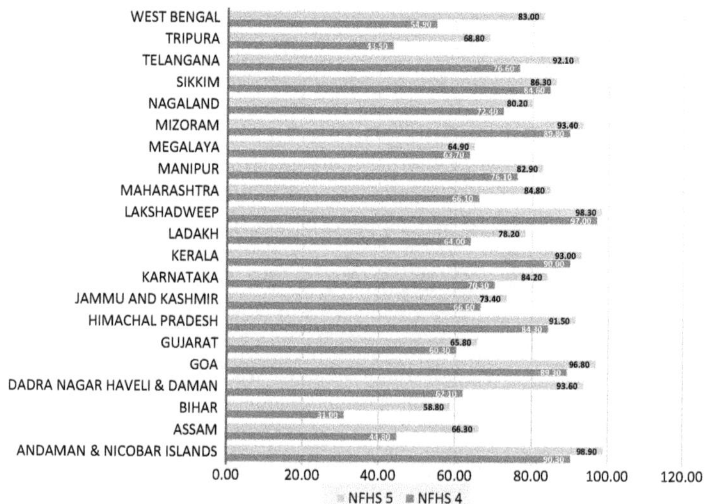

Figure 1.1 Percentage of Women in the Age Group of 15–49 Years using Hygienic Methods of Soaking Menstrual Blood across States of India: NFHS-4 and NFHS-5.

Sources: Compiled from (Sengupta et al., 2020). More women are opting for safe menstrual practices, NFHS-5 data shows. *Down to Earth, December 23.*

overview of the percentage change in adoption of hygienic methods to soak menstrual blood from NFHS-4 to NFHS-5. Figure 1.2 gives a rural-urban comparison for the same, as per NFHS-5 survey results.

The largest switch over from relatively less hygienic to hygienic menstrual products can be seen in Dadra Nagar Havel and Daman, a change of 31.5 percentage points. This is followed by West Bengal (28.1 percentage points) and Bihar (27.8 percentage points), as compared to NFHS-4.

Figure 1.2 shows that the percentage of adoption of hygienic methods of soaking menstrual blood is higher for rural areas in comparison with urban areas for Andaman and Nicobar Islands, Dadra Nagar Haveli and Daman, Goa and Lakshadweep. Among the other states, where the proportion of women adopting hygiene practices is higher in urban areas compared to the rural areas, the largest gap is in Meghalaya (25.9 percentage points).

India has made a leapfrog progress in terms of providing access to sanitation facilities to 100 per cent households. In 2019, India was declared open-defecation free (Verma, 2019). Adequate sanitation facilities are a prerequisite to menstrual hygiene management.

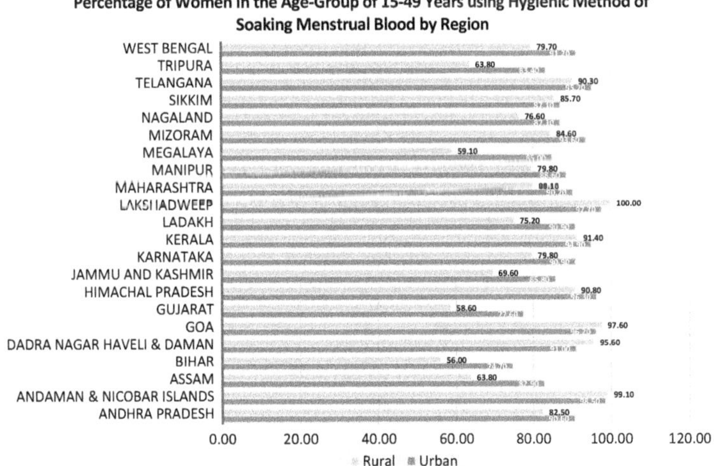

Figure 1.2 Percentage of Women in the Age Group of 15–49 Years using Hygienic Methods of Soaking Menstrual Blood across States of India: Rural and Urban (NFHS-5).

Source: Compiled from (Sengupta et al., 2020). More women are opting for safe menstrual practices, NFHS-5 data shows. *Down to Earth, December 23.*

1.2 Rationale for the Study

We have already discussed that menstruation is a public health issue (WHO, 2022) and that it has started gaining the attention of policy makers in past few years. Non-governmental organisations (NGOs) have been active for quite some time in spreading awareness about menstruation and supplying menstrual hygiene products, mainly sanitary napkins, either at subsidised rates or for free. Government has also made available menstrual hygiene products at subsidised rates. Adherence to hygiene involves costs, which are required to be borne either by an individual, organisation, or the state. Therefore, a methodology of estimation of costs for ensuring menstrual hygiene needs to be developed, and a wide variety of costs are required to be estimated. The cost estimation would give insights into an individual's affordability for the hygiene product. This estimation can also give insights to the policymakers about pricing the product (at subsidised prices) or making it available for free to women from lower socioeconomic strata.

Secondly, non-adherence to hygiene can result in infections and diseases, and menstruation is no exception. Failure to ensure menstrual hygiene may result in itching or infection in private parts (Brander, 1942; Cardwell, 1942; Durr-e-Nayab, 2005), and in extreme cases, it could even cause abscess. These ailments require treatment and would result in implicit costs of medical consultation, medicines, and, in extreme cases, even hospitalisation and surgery. Such associated illnesses can have an influence on absenteeism, which in turn, results in the loss of wages. The wages lost because of ailments resulting from failure to adhere to menstrual hygiene indicate the opportunity cost of menstrual hygiene. While implicit costs of medical care would be borne partly by the individual and partly by the government (if treatment is availed at the public healthcare facilities), the wage loss due to absenteeism is entirely borne by the individual. Therefore, we also attempt to identify the implicit cost components; the data collected through our survey did not have adequate responses to estimate implicit costs. Bhatia and Cleland (1995), quoted in the NIPCCD (2014) report, show that women have poor personal hygiene and are not able to maintain good sanitary conditions, mainly due to lack of awareness. They also find that women with poor personal hygiene and not having access to good sanitary facilities are more likely to have gynaecological symptoms, high-risk deliveries, and symptoms of Reproductive Tract Infections (RTIs). This also requires examining the extent of awareness about menstruation and menstrual hygiene among women across different socioeconomic strata.

Unhygienic sanitary conditions as per NIPCCD (2014) include:

Unclean sanitary pads / materials: they result in bacterial proliferation, which might be the cause of local infections, that may extend to the vagina and the uterine cavity.

Unsafe disposal of used sanitary materials or blood: they result in the risk of infecting others, especially with Hepatitis B.

Frequent douching (forcing liquid into the vagina): this can cause bacterial transmission and proliferation into the uterine cavity.

Not washing hands after changing sanitary napkins/cloth, etc.: this can result in developing and spreading the infections.

Menstrual morbidities are common among women. Garg et al. (2001), as quoted in NIPCCD (2014) show that out of 301 women from an urban slum in Delhi who underwent the clinical examination, a high prevalence of major reproductive infections was revealed: bacterial vaginosis (41.5 per cent), chlamydia (28.7 per cent), candidiasis (18.6 per cent), trichomoniasis (4.3 per cent), syphilis (4.2 per cent), hepatitis B (5.8 per cent) and hepatitis C (1.8 per cent). They also found that these 301 women, who were screened for four infections (bacterial vaginosis, candidiasis, trichomoniasis, and gonorrhoea), 56 per cent of them were found infected with at least any one of these. Ganie et al. (2019) observe that the prevalence of polycystic ovarian syndrome (PCOS) ranges from 3.7 per cent to 22.5 per cent in India, whereas the prevalence of thyroid is one among eight women (Velayutham et al., 2015). PCOS and thyroid can adversely affect the menstrual health, and therefore, it can have a multiplier effect on medical costs required to be borne by the individual or by the Government health centres/hospitals. There are relatively a fewer studies that provide conclusive evidence on the lack of hygiene or improper use of absorbents resulting in infections and rashes, though Cardwell (1942); Durr-e-Nayab (2005) and Mulgaonkar (1996) have observations to support this. Women are found to experience some health problems, like headache, stomachache, body ache, bloating, etc., or even psychological symptoms like anxiety, depression, mood swings, etc., a couple of days prior to menstruation, known as pre-menstrual syndrome (PMS). They also experience these symptoms during their menstrual cycles. Sometimes, they require a doctor's consultation and take medicines for these pre-menstrual syndromes and menstrual problems. This too adds to the cost of menstruation. These are indirect costs of menstruation, *albeit* explicit costs.

Thirdly, little work is undertaken to estimate the direct costs of menstrual hygiene. Costs could be one of the factors that might prevent a girl/woman from following hygienic practices. Hygienic practices do not mean using a sanitary napkin or a menstrual cup. Even reusable cloths could be hygienic if they are changed at regular intervals, washed with a good detergent, dried in sun, and avoided using damp cloths, constitututing good hygienic practices. It is likely that certain sections of the society might not be able to afford a washing detergent. Even those who use sanitary napkins or tampons require changing them at regular intervals. Thus, irrespective of the absorbent used, practices could be hygienic or unhygienic, and that may imply explicit (and implicit costs) associated with practicing (or otherwise) menstrual hygiene. Therefore, the present study examines the awareness of women about menstruation and menstrual hygiene, beliefs held, and taboos practiced by them about menstruation. This study further examines the menstruation practices among women in Gujarat, which include the type of absorbent used, frequency of changing the absorbent, the method of disposal/cleaning, handwashing habits during menstruation, the habit of washing the private parts during menstruation, etc. All these help us in estimating direct and explicit costs of menstrual hygiene.

Regressive norms and beliefs associated with menstruation, result in taboos. Menstruation is therefore not only a biological phenomenon, but also has its roots in anthropology, which in turn, emanates as culture, and social norms (Sridhar, 2019) and has implications for beliefs, taboos, and practices. Beliefs and taboos come in the way of practicing hygiene, whereas lack of awareness contributes to reduced hygiene. Therefore, beliefs and taboos associated with menstruation also require to be understood to get insights into the sociocultural aspects that influence the hygienic practices.

The objectives of this study are:

1 To examine the extent of awareness, beliefs, and taboos associated with menstruation.
2 To elicit patterns in menstrual hygiene practices among women.
3 To understand the problems faced by women because of lack of menstrual hygiene.
4 To estimate the explicit and implicit costs of menstruation and menstrual hygiene.

The present study examines the menstrual hygiene awareness and practices among women in Gujarat. The study also examines the cost components and attempts to estimate the explicit and implicit costs of menstruation. This is because people belonging to the lower socioeconomic strata might have to spend a significant portion of their monthly income to ensure hygiene or for medical expenses for the ailments resulting from poor hygiene, or even bear the burden of wage loss due to absenteeism.

The focus on examining the components of explicit and implicit costs remains largely confined to individual costs. Though, this study makes a brief remark on the potential impact on environment, resulting from the disposal of menstrual waste.

1.3 Methodology of the Study

This study makes use of mixed methods – qualitative and quantitative – for triangulation purpose. Since this study had to be undertaken during the first lockdown phase of COVID-19, adaptations had to be made to minimise the disruption in the data collection process. The details of adaptation are also discussed here.

1.3.1 The Sampling Frame

The sampling frame for this study consists of menstruating women in the age of 14–49 years. It may be noted that we have not included transgenders (who were assigned female sex at birth) in our study. The average age of menarche is 13.76 years (Pathak et al., 2014), and the average age of menopause is 46.2 years for Indian women, whereas it is 51 years for the women from the western countries (Ahuja, 2016). It is imperative to state here that though the age range of women was 14–49 years, the sampling frame included only those women who were menstruating in this age group, when the survey was undertaken. This means that women who were pregnant at the time of the survey or were already in menopause or had menstrual dysfunction or girls who had not experienced menarche were excluded from the sampling frame. This is because the objective of the study is to get insights into practices, explicit costs, implicit costs, and opportunity costs. Therefore, the data on current practices and costs would be useful for making a meaningful comparison across various socioeconomic strata. The details of practices and allied costs for the pregnant or menopausal women would not give a comparable estimate with those who currently menstruate. Moreover, menopausal women and those with menstrual dysfunction would have different sets of health issues, which would not be comparable with those of menstruating women. The question of practices and allied costs doesn't arise for the girls who have not yet entered their menarche. Therefore, the sampling frame consisted of women in the age of 14–49 years and those who were in their menstrual phase at the time of the survey.

Gujarat is a coastal state and shares an international border with Pakistan on its northern boundary. Gujarat has one of the strongest economic fundamentals among the states of India; it contributes 7.99 per cent share to India's GDP, 16.8 per cent share of industrial output (the highest in India), with just 4.99 per cent population of the country. Gujarat also has the lowest unemployment rate (Government of Gujarat, 2020).

1.3.2 The Design of Survey Instruments

The survey was undertaken in three phases over a period of six months: (1) In-depth interviews with gynaecologists and women doctors (2) administering questionnaire on the respondents sampled from the sampling frame, and (3) in-depth interviews of women and Accredited Social

Health Activists (ASHA) to get deeper insights into the hygiene practices and the reasons for adopting the same, and to triangulate the findings derived from the data collected through the questionnaire.

1.3.2.1 Interview Guide for Gynaecologists/Women Doctors

Based on the study objectives, the gynaecologists were asked to provide us with the information about the problems associated with menstruation or during menstruation, faced by women. The doctors were asked to give us information on what parameters determine whether the practice is hygienic or not, whether the patients examined by them adopt hygienic practices or not, how frequently they visit the doctor, and the approximate expenditure on menstrual health borne by the patients, including medicines/tests, etc. The doctors were also asked whether their patients discuss their menstrual health problems without any hesitation or if the doctors have to put in efforts to understand their problem. We sought information on the broad socioeconomic background of their patients who sought consultation on menstrual/vaginal health. We did not seek socioeconomic profile of individual patients because the objective of these interviews was to strengthen the understanding of menstrual health issues and whether they differ across regions or any other socioeconomic characteristics. In-depth interviews with one gynaecologist, one Ayurvedic doctor specialised in women's health, one male gynaecologist, and a doctor from the government hospital of The Dangs (a tribal district) were conducted to get insights about their opinion about the awareness about menstruation and menstrual hygiene practices adopted by women in tribal areas. The interview guide for conducting in-depth interviews of gynaecologists/women doctors is given in Appendix – A.

1.3.2.2 Questionnaire Design

Questionnaire design was a longer exercise and was primarily based on the responses received from the gynaecologists/women doctors. The insights derived from reviewing the literature (Chapter 2) are also incorporated in the questionnaire design. We also had Public and Patient Involvement consultations for the questionnaire design. Working and non-working women who participated in the PPI have commented on the adequacy of different issues and suggested modifications to the questions and responses. We incorporated those suggestions, and were further discussed by independent reviewers appointed by the Indian Council of Social Science Research (ICSSR), the organisation that funded this study. We also incorporated the comments and suggestions given by them. The questionnaire was then translated into Gujarati and pilot tested. The issues faced during the pilot testing of the questionnaire were rectified, and the revised questionnaire was retested for accuracy and understanding. After three major and two minor modifications, the Gujarati as well as the English versions of the questionnaire were administered to the respondents. The final (English) version of the questionnaire[5] is provided in Appendix – B.

1.3.2.3 The Questionnaire

The questionnaire is designed to include questions on (1) awareness, beliefs, and taboos (2) practices during menstruation, and (3) explicit and implicit costs of menstrual hygiene. The questionnaire is organised into four sections. First section is designed to capture the socioeconomic profile of the respondents. Section 2 has three subsections on awareness, beliefs, and taboos. Third module seeks responses on the hygiene practices adopted by them during menstruation. Fourth module seeks information on the explicit and implicit costs associated.

1.3.2.4 In-Depth Interviews of Women

After collecting the responses using the questionnaire, in-depth interviews were conducted with women, to get deeper insights into the problems associated with menstruation and menstrual hygiene. The interviews focused mainly on physical and psychological problems faced by women during menstruation.

In-depth interviews focused on issues faced by women during their entire span of menstruation since menarche, remedies sought (or otherwise), whether they were practicing home remedies or they consulted a doctor, what was their level of understanding, etc. In these interviews, the women were largely asked to narrate their experiences about menarche, issues faced by them during menstruation, and their coping strategies for psychological problems and mood swings. The guide for in-depth interviews is given in Appendix – C.

1.3.3 The Survey Exercise: Planned versus Adapted to COVID-19 Pandemic Situation

The survey was originally planned to be conducted in five districts of Gujarat. The selection of the districts was made on the basis of their level of development. Thus, two developing districts (of which, one had to be a tribal district), two developed districts, and Surat district (the place where this research is based, and it is also a developed district) were chosen for the survey. The share of urban population in a district is used to classify it as a developed or developing district. There is evidence of a positive relation between urbanisation and economic growth (Chenery and Taylor,

1968; Henderson, 2003). Hence, percentage of the urban population is considered to be an indicator of the level of development of a district. The endogenous growth models of Lucas Jr (1988) and Romer (1986) theoretically explain the positive association of urbanisation with knowledge spillovers, over and above productivity. Hence, it may be argued that a woman from an urban area is likely to have more information about menstrual hygiene and more access to menstrual products as compared to those from rural areas. This further justifies the stratification for district selection based on the extent of development (urbanisation).

Research design and sampling procedures, among other factors, influence the external validity of a study. External validity broadly refers to the extent of generalisability of research findings, though one may find minor variations across the research literature on definitions of external validity (Lucas, 2003). Therefore, multi-stage stratified sampling was proposed (pre-COVID 19) to be adopted to deal with the issue of external validity of this study.

A sample of 150 women from each of the five districts was planned to be collected, which would work out to a total of 750 women for five districts. However, all the plans went topsy-turvy, following the pandemic and subsequent imposition of lockdown in the entire country. When the lockdown was imposed in the end of March 2020, pilot testing of the questionnaire was already completed. While the lockdown was greatly relaxed after three months, it was soon realised that normalcy of life would take much longer to get restored. Therefore, the survey was adapted to online and telephonic interviewing in order to minimise the disruption in the study and ensure the safety of the field investigators and respondents.

Since we switched over from face-to-face surveys to online and telephonic surveys, alternative methods of undertaking the survey had to be considered in order to ensure the least disruption of the survey work. Sampling method also had to be changed to remotely contact the respondents. The impact of online surveys and lack of personal communication was observed in the form of very low response rates, high rate of item-specific non-response, and a compromised quality of responses to the questionnaire. Therefore, further adaptation had to be worked out to collect quantitative data. Qualitative data were to be collected through in-depth interviews and focus group discussions, which also had to be adapted to be conducted over online meeting platforms like Zoom.

ILO (2020) describes at length, the COVID-19 pandemic's impact on survey operations and suggests responses to the challenges faced. A lot of countries use Pen and Paper Interviewing (PAPI), Computer-Assisted Personal Interviewing (CAPI), and Computer-Assisted Telephonic Interviewing (CATI). Though a few countries could switch over from PAPI and CAPI to Computer-Assisted Web Interviewing (CAWI), there were disruptions in surveys for most of the countries that couldn't adapt to CAWI. Thus, switching over to CAWI and CATI minimised the disruptions in the survey operations for our study.

1.3.3.1 Adaptation of Sampling Method

Sampling procedures have implications for the selection of analytical tools and also on generalisation of results. This is an exploratory study, and no attempt is made to generalise the findings. The insights derived from this study could be helpful to replicate the study across geographical regions, over time and also plan some interventional studies. We adapted our survey from multi-stage stratified sampling to respondent-driven sampling (RDS)[6], considering the circumstances, limited time frame for collecting the data, and the lack of information to enlist the population within the defined sampling frame. Probability-based sampling designs require a collectively exhaustive list of units to be sampled. Goodman (1961) recommends that the base sample (the first-stage respondents) be chosen using random sampling to induce the randomness required to apply the methods of statistical inference. However, first stage/first contact samples (hereafter, referred to as "first stage") were chosen on the basis of availability of WhatsApp, thereby resulting in a convenience first stage sample, as in the case of RDS.

1.3.3.2 Adapting to Online Self-Administered Questionnaires

The very first step towards adaptation to the COVID-19 situation was to develop an online version of the questionnaire and then pilot test it. Since the sample comprised women from different socioeconomic strata, the questionnaire was prepared both in Gujarati and English. The offline, printed questionnaires were in Gujarati only. Therefore, the offline Gujarati version was first translated to English, and then the online version was prepared. The original printed questionnaires were planned to be administered by the women project staff. It was originally planned that the women team members, along with the field investigators, would travel to the five sampled districts and get the questionnaires administered under their supervision. However, in the changed and challenging adverse circumstances due to COVID-19 situation, a decision to circulate the questionnaire link through WhatsApp was taken. The link was shared with the project staff, and they circulated it to their women contacts, fulfilling our sampling frame criteria, in Gujarat. The project team further requested their said contacts in Gujarat to further circulate the questionnaire link to their contacts in the same age range.

On one hand, where the randomness in selection had to be sacrificed because of snowballing, on the other hand, the geographical coverage widened to include many more districts of Gujarat vis-à-vis five as originally planned. However, this method of data collection had two major shortcomings: (1) the response rate reduced with the level of chain in the snowballing, and (2) the quality of responses was poor. De Leeuw (2012) has already cautioned of high non-response rate in internet surveys, over and above coverage and sampling biases. De Leeuw (2012); Frippiat et al. (2010); Couper (2000) have also cautioned about coverage bias in an internet survey. Coverage adequacy is compromised because of the digital divide. Therefore, an alternative method of data collection had to be thought of and subsequently adopted after pilot testing.

1.3.3.3 *Online Surveys Administered by Field Investigators*

Having understood the reasons for poor responses on online self-administered surveys, it was decided to employ women field investigators (FIs) and provide them training to administer the questionnaires over telephone/mobile phone. Though, some countries have made changes in questionnaire content for the ILO (2020) surveys for improving the response rate, and Bulsari et al. (2020) also suggest changes in questionnaire content to facilitate the administration of the survey. It was possible to retain the content of the questionnaire and improve the quality of responses by employing FIs. FIs were employed from those five districts that were originally planned to be selected for the study. These districts are: (1) Vadodara, (2) Anand, (3) Surat, (4) Tapi, and (5) Navsari. Educational background and relevant experience have a bearing on the quality of the survey work. Hence, those who possessed a post-graduate degree in disciplines of social sciences and also had some exposure to survey research were selected for the survey work; a few of them were pursuing their doctoral studies. Out of the 12 FIs employed for this survey, five belonged to the rural areas, whereas seven belonged to the urban areas.

Vadodara and Anand are developed districts, whereas Tapi and Navsari are tribal districts, and Surat is the host district for this study. Thus, we could get back quite close to our originally planned design of multi-stage stratified sampling, where the first stage stratum was districts and the second stage stratum was the region (rural/urban). In each stratum, women belonging to different age groups, socioeconomic strata, and levels of education were selected, though convenience sampling was adopted in the third level.

FIs were given a rigorous telephonic training on a one-to-one basis over WhatsApp video calls. Videos were recorded, explaining each question in detail, and shared privately with the FIs through YouTube.

FIs were asked to identify the respondents for the survey in their surroundings to begin with. Once all the references in the vicinity of their residence were exhausted, they could expand the respondents' base by exploring the references received by them from the respondents whom they already surveyed. They were given a choice to either do a face-to-face survey or do a telephonic/mobile phone (or WhatsApp call) survey.

Initially, three FIs were asked to begin the primary survey data collection and share their experiences. On receiving a positive response for face-to-face as well as telephonic surveys from these three FIs, the other nine were asked to begin their survey work. It was observed that both – the respondents and a majority of FIs preferred face-to-face interaction rather than a telephonic interview/conversation. However, soon after the lockdown was relaxed and during the local peaks of COVID-19, respondents preferred telephonic conversation over face-to-face surveys.

Ethical guidelines included explaining the objective and purpose of the study to the respondent, assuring them of confidentiality and anonymity, and not forcing her to complete the survey (letting her leave the survey at any juncture she feels uncomfortable; *albeit*, making her feel comfortable throughout the survey was also explained in the training). Explicit written consent was sought from respondents before beginning the survey.

The response rate of questionnaires administered by FIs was close to 100 per cent. Most of the respondents were first-level contacts of the FIs. Further referrals were not required as the FIs could approach the targeted sample size. Each FI was given a target to administer 50 questionnaires.

The quality of responses was far superior to those of self-administered questionnaires.

1.3.3.4 Offline Self-Administered Surveys

The questionnaire copies were printed for pilot testing before the lockdown. Thus, quite a few questionnaires were distributed to get responses in the Surat and Vadodara districts soon after the lockdown was partially relaxed. The response rate of these offline questionnaires was 100 per cent.

However, the quality of responses of offline and online self-administered questionnaires was equally poor. This could be because it is difficult to ensure internal consistency across the respondents. The interpretation of a question or its fixed-choice alternative responses would differ across the participants depending upon their sociocultural background, level of education, and exposure to other societies and cultures.

The response rates as well as the quality of open-ended questions were better compared to their self-administered online counterpart. This could be because it is easier to write long sentences as compared to typing them on a smartphone or a tablet. Moreover, psychologically, one may tend to see more space on a paper sheet as compared to that on a text box on a smartphone screen.

While studies (Ahmed et al., 2018; Marcano Belisario et al., 2015) reveal that deploying Apps would improve the quality of responses in comparison with offline self-administered questionnaires, there is a dearth of literature on comparison of paper-based versus online questionnaires. Moreover, these studies do not provide details on the quality of open-ended versus close-ended questions' responses or on item non-responses.

1.3.4 In-depth Interviews of Women and NGOs

Phase – III of this research aimed at getting deeper insights into menstrual hygiene awareness, with an objective to get insights into the costs of maintaining hygiene/costs resulting out of an inability to maintain hygiene during menstrual cycles.

In-depth interviews were planned to be conducted for women with menstrual health conditions or those requiring frequent consultations with a doctor/gynaecologist. This was originally planned to be done by extracting information from the responses of questionnaire. However, because of the disruption of the schedule of Phase – II of administering the questionnaires, Phase – III was simultaneously implemented using a snowball sampling. In-depth interviews were conducted using both – face-to-face and telephonic modes. The mode of interview – face-to-face vis-à-vis telephonic – had no influence on the quality of the responses. The depth of the response depended more on the complexity of the menstrual health issues experienced by them, respondents' self-awareness, and their education. We conducted 16 in-depth interviews.

In-depth interviews for NGOs were done telephonically, with two NGOs: one working in the local area of a big city in Gujarat, and another Gujarat-based NGO, working nationwide. The interview guide for NGOs is given in Appendix – D.

Focus group discussions (FGDs) were conducted to get deeper insights and to validate the outcomes of the data collected using a questionnaire. FGDs were conducted on a homogeneous group, as well as on a heterogeneous group. Homogeneity/heterogeneity was determined on the basis of age, level of education, marital status, and other socioeconomic characteristics. Homogeneous groups had seven participants, and heterogeneous had five of them. FGDs were conducted virtually using Zoom Meeting; *inter alia*, these could be conducted only with those who had access to technology and were literate.

1.3.5 Conceptual Framework

Menstruation is inevitable to a woman's major life span. Despite that, it continues to face the social taboos. Age-old beliefs and taboos associated with menstruation have resulted in poor awareness about menstruation among women. This makes menstrual hygiene management (MHM) difficult. MHM is important because menstruation is a health issue, and menstrual cycles occur every 28 days. Menstrual health is necessary not only for ensuring good reproductive health but also for ensuring overall good health. There seems to be a dearth of literature on the costs of MHM. The direct costs of MHM include the use of menstrual products as well as products used to ensure hygiene in private parts. The cost of soap/disinfectant to clean the hands after using the washroom also constitutes direct costs of MHM. Indirect costs include medical costs to treat PMSs and aliments during menstruation. In many societies, women are not allowed to do some or all household chores during their menstrual cycles. They might need to employ paid help when other family members are not able/available to do those household chores. This too is an indirect cost component of MHM.

Implicit costs include wage loss due to absenteeism and medical expenses incurred to treat the ailments that might have been developed because of their inability to ensure hygiene during menstrual cycles. It is difficult to estimate the opportunity cost for absenteeism because of its possible long-term effect on careers, promotions, etc. Estimating the opportunity cost of school/college absenteeism is also difficult, though we attempt to estimate the opportunity cost of absenteeism for working women.

We have used a bottom-up approach for cost estimation. The conceptual framework for this study is given in Figure 1.3:

The discussion so far does not take into consideration the women with psychiatric disorders. Browne et al. (1989) show that the expression, understanding, and treating of women with psychological disorders, resulting in communication problems, is challenging. Therefore, in our study, we have confined only to women with no psychological disorders.

1.4 Sample Distribution and Respondents' Profile

Sample selection, along with many other factors, plays an important role in strengthening the external validity of the study. We have already discussed the sampling approach and adaptation to it in Section 1.3 1. Finally, we had the representation of women respondents from 24 districts of Gujarat. The distribution of responses by districts, categoriszed by rural and urban areas, is given in Appendix – E (Table 1.1). In the original, pre-COVID-19 plan, district selection required choosing two developed districts, two developing districts (of which one would be tribal), and Surat. The distribution of the sample across developed, developing, and tribal districts is given in Table 1.1.

It can be seen from Table 1.1 that the sample is almost equally distributed across rural (46.05 per cent) and urban (53.95 per cent) regions.

We got an overwhelming response to this survey. We collected data from 1025 women, more than 35 per cent of what was planned. All the analysis is undertaken using R software, and graphs are generated in MS-Excel. The patterns in educational attainment, occupation type, household income category, house type, access to toilet facilities, marital status, category[7], religion, and food habits of these 1025 respondents can be found in Figures 1.4–1.6:

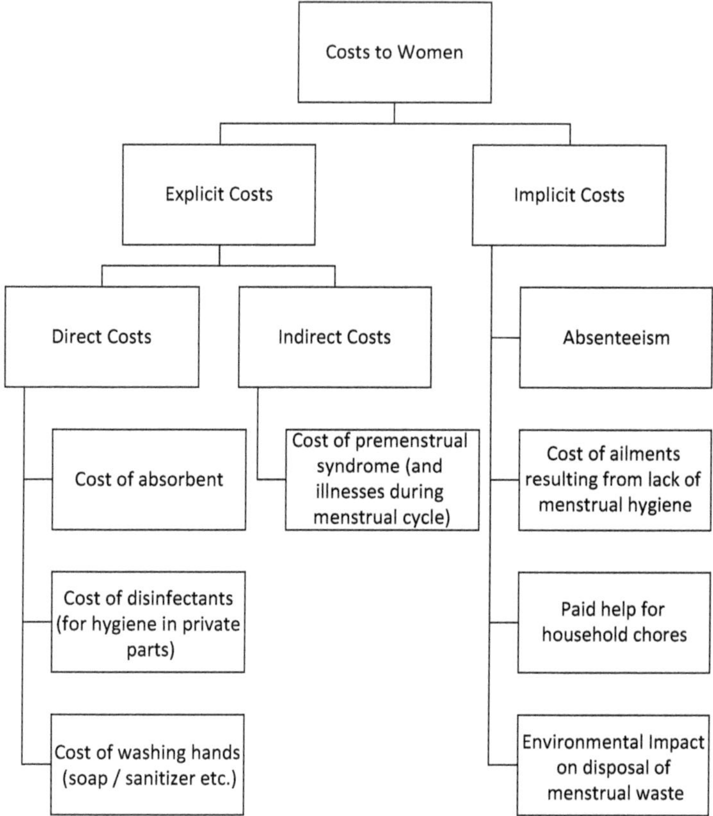

Figure 1.3 Conceptual Framework for Estimating the Cost of Menstrual Hygiene for Women.

Figure 1.4 shows that a majority of respondents (66.15 per cent) have completed Higher Education. There is less representation of respondents with lower levels of education. It may be noted that many girls might still be pursuing their education. The level of education is an indicator of their ongoing education level at the time of survey.

Nature of occupation might also influence menstrual hygiene practices. Availability and accessibility to sanitation facilities at workplace would influence hygiene practices. Also, the working hours, and nature of work (office work, manual labour, etc.) would also influence hygiene practices during menstruation. The data on occupation shows that a majority of the respondents (24.39 per cent) are college-going students, followed by 23.71 per cent, who work indoors. The representation of school-going

Table 1.1 Sample Distribution across Development Status of the District and Region

Regions		Rural	Urban	Total
Developed	Count	260	427	687
	Row Percentage	37.85	62.15	100
Developing	Count	91	103	194
	Row Percentage	46.91	53.09	100
Tribal	Count	116	21	137
	Row Percentage	84.67	15.33	100
Unclassified	Count	5	2	7
	Row Percentage	71.43	28.57	100
Total	Count	472	553	1025
	Row Percentage	46.05	53.95	100

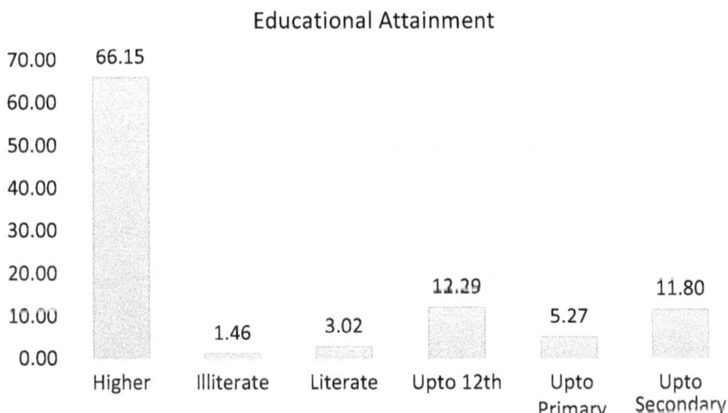

Figure 1.4 Educational Attainment of Respondents.

girls and women in manual labour work is quite low (5.76 per cent and 4.59 per cent respectively) (Figure 1.5).

In the pre-pandemic days, when the questionnaire was pilot tested, the respondents were asked to give their actual monthly household income. However, considering the pandemic situation and issues faced while receiving the responses through the online self-administered question-naires (as discussed in Chapter 4), it was decided to categorise household income of respondents into high, low, lower-middle, and upper-middle categories. This categorisation is based on the World Bank's country clas-sification by their Gross National Income (GNI) per capita[8]. These GNI per capita are given in US dollars. To convert the US dollars into Indian Rupees, the Rupee-Dollar conversion rate of Rs 75 (average for 2019–20) is taken. A majority of the respondents (46.44 per cent) belong to

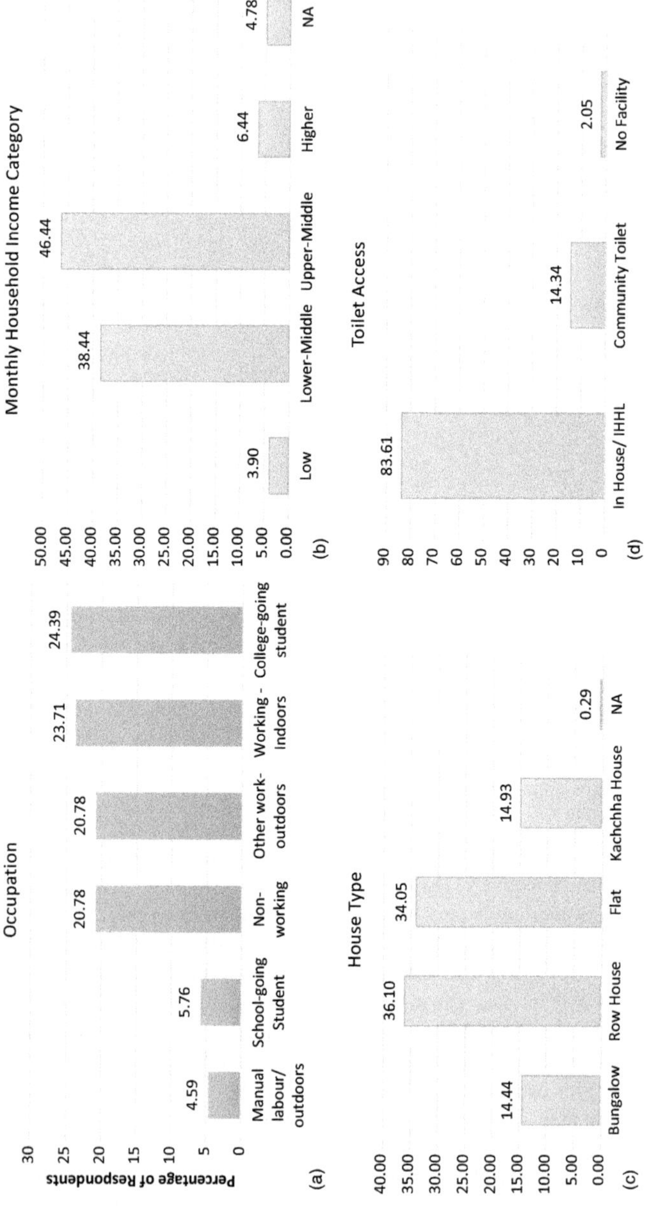

Figure 1.5 Socioeconomic Characteristics of Respondents: Occupation, Income, House Type, and Access to Toilet Facilities.

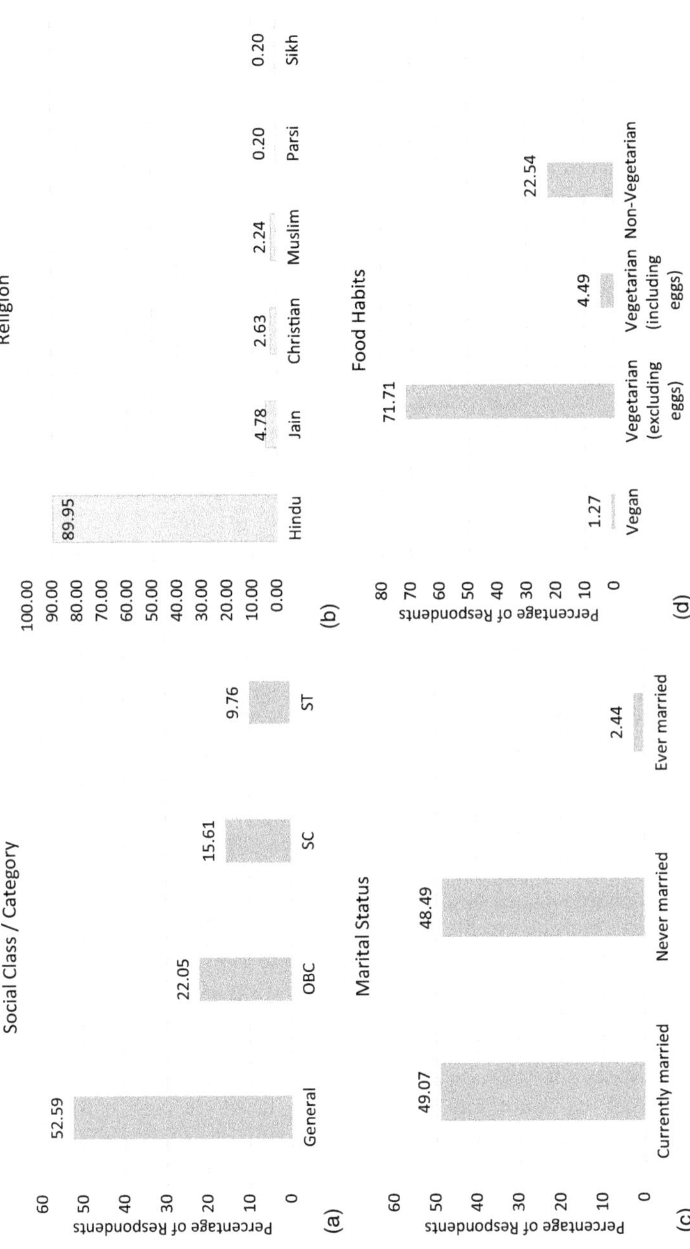

Figure 1.6 Sociocultural Characteristics of Respondents: Category, Religion, and Food Habits.

Note (to Figure 1.6(d)): Eggs are not a part of vegetarian diet in Indian context and therefore, we have split vegetarian diet in two categories. Fish and seafood is clearly a part of non-vegetarian diet in the Indian context.

upper-middle-income category. Representation of women in low-income and high-income categories is relatively less (3.90 per cent and high 6.44 per cent respectively). Also, 4.78 per cent respondents have chosen not to give response about their incomes. Income is also hypothesised to influence menstrual hygiene practices and purchasing power to buy related products.

In order to get a clearer idea of the economic status of the respondents and to cross-examine the income reported by respondents, questions on the type of house are asked. Majority respondents (36.1 per cent) are found to live in row houses, whereas 14.93 per cent live in "kachcha" houses[9].

Access to sanitation facilities is important to ensure menstrual hygiene. Therefore, information access to toilet facilities was also sought. Out of 1025 respondents, a majority (83.61 per cent) of respondents have access to toilets within their house premises, 14.34 per cent respondents have access to community toilets. However, 2.05 per cent respondents have no access to a sanitary toilet.

The study hypothesises that the category to which a respondent belongs would have no effect on awareness or practices. However, taboos may significantly differ for the tribals, that is, for those belonging to Scheduled Tribe (ST) category, as compared to other categories. Here, 52.59 per cent respondents belong to the General Category, followed by 22.05 per cent, who belong to OBC, whereas 15.61 per cent belong to Scheduled Caste (SC). However, only 9.76 per cent respondents belong to ST category. Despite a low proportion of ST respondents, the absolute number of ST respondents is sufficient to examine the differences in taboos and, thereby, on practices.

Beliefs about menstruation differ across religions. They are reflected in the form of taboos and practices associated with menstrual hygiene. Therefore, the information on religion of the respondent was also sought. Out of 1025, a majority of respondents (89.95 per cent) are Hindus.

The marital status of a woman might influence the menstrual hygiene practices. The chances of adopting hygienic practices by those who have children are higher because of their exposure to hygienic practices at the time of childbirth and just after that. There is almost equal representation of married and single women/girls.

It is believed that the type of food consumed during and around menstrual cycles affects the onset of menstrual cycles as well as influences the pre-menstrual syndromes and problems faced during menstruation. Therefore, the respondents were asked to specify their food habits. Out of 1025 respondents, majority (71.71 per cent) are vegetarian.

Respondent characteristics like their age at the time of survey, age at first marriage, number of family members, number of children, etc., add to the sociocultural description of the respondent. The number of rooms in the house, in addition to the income category and house type (Figure 1.5) helps in getting information about the socioeconomic condition of the respondents. These characteristics are measured on a continuous scale, and hence, they are summarised separately, as in Table 1.2.

Despite that, the originally planned sample including women in the age range of 14–49 years, two girls of age 12 and 13, who have attained menarche, and one 52-year-old woman, who has not attained menopause, have responded in the self-administered questionnaires, and hence, they are retained for the purpose of further analysis. The mean age of the respondents is 27.73 years.

The number of family members ranges from a minimum of one member to a maximum of 22 members. The average family size is of 5 members. The distribution of number of male and female members in a family reveals that there are households ranging from only women households to households with a maximum of 12 male members. The maximum number of female family members is found to be 10. The median male and female members in families are 2 each.

Table 1.2 Age and Family Members Summary Statistics

	Min.	1st Qu	Median	Mean	3rd Qu	Max.	N (Sample Size)
Age	12.00	22.00	26.00	27.73	33.00	52.00	1025
Age at First Marriage	14.00	20.00	23.00	22.33	25.00	32.00	537
Number of Children	0.000	1.000	1.000	1.438	2.000	5.000	535
Number Of Family Members	1.000	4.000	5.000	5.026	6.000	22.000	1025
Male Family Members	0.000	2.000	2.000	2.269	3.000	12.000	1025
Female Family Members	1.000	2.000	2.000	2.717	3.000	10.000	1025
Number of Rooms (Excluding Kitchen)	1.000	2.000	3.000	2.961	4.000	12.000	1025
Age at Menarche	9.00	13.00	14.00	14.48	16.00	28.00	1025

Note: The last column of Table 1.2 gives the number of respondents, whose data are used to estimate the summary statistics. The data are collected from 1025 women, out of which, 537 are either currently married or were married but not with their partner now. Thus, 537 women have given their age at first marriage. Out of 537 married women, 535 have (or given the data on) number of children.

Out of 51.51 per cent women who are either currently married or were married (but not currently with their husbands), the mean marriage age was 22.33 years. The age at their first marriage ranges from as low as 14 to as high as 32 years. The median number of children of these women who are currently married or were ever married at some point in their lives is one. The number of children for these married women ranges from zero (minimum) to five (maximum).

The number of rooms in the house in which these women live, ranges from one (minimum) to 12 (maximum). The median number of rooms in the house of these respondents is three.

Age at menarche ranges from nine to 28 years of age. The median age of menarche is 14 years.

The respondents were asked to mention whether they, or any of their family members, have any addictions, like smoking, chewing tobacco, alcohol, etc. It is likely that the respondent or her family member might have more than one addiction, for example, smoking and alcohol. Therefore, frequencies and percentages are calculated while considering multiple responses to the addiction. Figure 1.7 shows the patterns of addictions among respondents and their family members. It can be seen that more than 99 per cent of respondents and more than 90 per cent of their family members have no addiction. This data is collected to explore whether there is any association between addictions and problems faced prior to/during menstruation, or whether family members' addictions, like smoking, have any influence on menarche.

It may be noted that since there are a few respondents who have more than one addiction, and the same is true for their family members, the

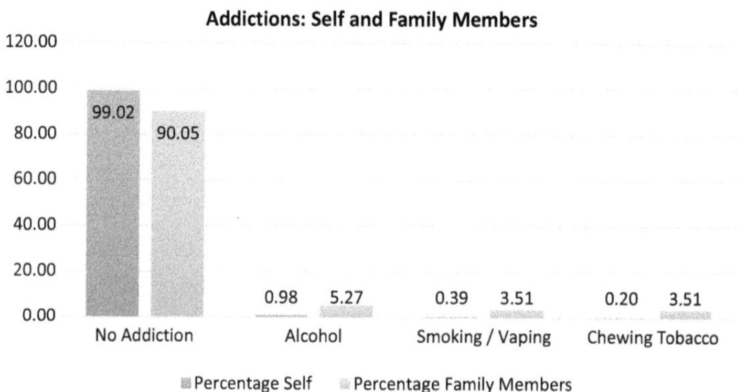

Figure 1.7 Addictions – Self and Family Members.

sum total of percentage exceeds 100 per cent. Also, the respondents were asked whether any of them/their family members are addicted to drugs. The data reveals that not a single respondent or any of her family members are addicted to drugs.

1.5 Chapter Scheme

This book is organised in five chapters:

Chapter 2: Menstruation Awareness, Beliefs, and Taboos: The chapter begins with a discussion on the existing literature on awareness and practices about menstrual hygiene. The patterns of awareness about menstruation are presented, and the awareness score is calculated to examine the impact of socioeconomic factors on awareness. The influence of socioeconomic factors on awareness score is examined using a regression model, and the results are presented and discussed. This is followed by a discussion on beliefs and taboos in the available literature. Patterns in beliefs about menstruation are then discussed, and a belief score is calculated. The influence of socioeconomic factors on belief scores is examined using a regression model. A similar discussion on patterns of menstruation taboos, and the influence of socioeconomic factors on the taboos score follows. The chapter concludes with a summary of these findings.

Chapter 3: Menstruation Hygiene Practices and Direct Costs: This chapter begins with a discussion on the patterns in menstrual hygiene product awareness and the usage pattern of menstrual products. Then, a discussion on the influence of socioeconomic factors on the choice of menstrual products follows and is based on the results of a logistic regression model. While menstruation awareness is already discussed in Chapter 2, this chapter does discuss the extent of product awareness before proceeding to the discussion on usage patterns. This is followed by a discussion on product preferences and the preferred and usual place from where the product is purchased, whether the product is bought by themselves (or somebody else buys it for them), whether they experience any hesitation in asking for menstrual products, the person accompanying the respondent while she goes to purchase these products, etc. These factors may influence the product preference, which in turn, would influence the direct cost of menstrual hygiene, and hence, responses were sought on these questions. A detailed description of the methodology to estimate the per person annual direct cost of menstrual hygiene is followed by the annual estimate of the direct cost of menstrual hygiene. The chapter ends with summarising the results on menstrual hygiene practices and by presenting the yearly estimate of the use of disposable sanitary napkins/tampons, and of using menstrual cups (a reusable menstrual product).

Chapter 4: Indirect and Implicit Cost Components of Menstruation: This chapter begins with a discussion on the prevalence of PMS and ailments experienced by women during menstruation. This is followed by a description of the methodology for estimating the medical costs of PMS and menstruation ailments (indirect costs of menstruation) and providing the annual cost estimate for the same. This is followed by a description of the estimation of annual costs of doing household chores like cooking, cleaning, etc., by employing paid help, apportioned to the days of the menstrual cycle. The underlying issues in estimating such costs are also discussed in this chapter.

Then, we discuss implicit costs of menstruation, which highlights the causes of absenteeism and depicts a pattern in the reasons for their absenteeism during their monthly menstrual cycles. Then, the methodology of estimating the annual opportunity cost of absenteeism for working women is described. Women suffer from different ailments in private parts because of failure to comply with menstrual hygiene. This also involves medical expenses. A description of the pattern on the prevalence of such ailments and a discussion on the methodology of estimation of the annual medical costs for these ailments follow. There is a high prevalence of medical conditions like PCOS, thyroid, etc., among women, which has implications on regularity and consistency of menstrual cycles. The methodology of estimating yearly medical costs to treat these is described in this section. Women also experience menstrual dysfunction. The presence of medical conditions or menstrual dysfunction has implications on psychological well-being of women. The implication of psychological well-being on medical costs is also discussed in this chapter. This study focuses on examining the implicit costs required to be borne by an individual woman. However, there is a brief discussion on the potential environmental costs because of inappropriate disposal of used menstrual product waste. The chapter ends with a summary that elicits and describes the indirect and implicit cost components of menstruation.

Chapter 5: Summary and Policy Implications: A summary of the study is presented in this chapter. This is followed by the highlighting the measures already undertaken by the Government to inculcate hygienic practices among women, to manage their monthly menstrual cycles. A discussion on the initiatives undertaken by various NGOs is also discussed in this chapter. Finally, the chapter concludes with suggestions for policy measures to create awareness among both women and sensitise men about menstruation with the objective of a healthier society.

Notes

1 It may be noted that not only women menstruate, but transgenders who were assigned "female" sex at birth also menstruate. Hence, the term "menstruators" is used, unless explicitly referred to by a study/ dataset as "women". Our study includes only women.
2 Wealth index is a composite measure indicating the living standard of a household.
3 Number of post-menarche girls in the age group of 15–19 was 120,410 during the NFHS-4 survey and they were 121,803 during the NFHS-5 survey. The respective surveys were conducted during 2015–2016 and 2019–2021. It would be imperative to note that NFHS-5 coincided with the COVID-19 pandemic and with our study too.
4 Others include women belonging to Jain, Buddhist, Sikh, and other religions.
5 Online administering of questionnaires had to be done, following the restrictions imposed because of the COVID-19 pandemic. The detailed discussion on the adaptation is given in Section 1.3.3 of this chapter.
6 Refer Handcock and Gile (2011) for RDS. There are many alternatives of RDS: network sampling and the more common, snowball sampling, each with minor differences in approach to the first stage samples (Coleman, 1958; Goodman, 1961; Granovetter, 1976).
7 There are four categories as per India's Affirmative Action Plan: Scheduled Castes (SC), and Scheduled Tribes (ST), Other Backward Classes (OBC), and Economically Weaker Section (EWS). Those not belonging to either of these are classified as "General". Unlike SC, ST, and OBC, which are largely social classes, EWS is an economic category. Moreover, EWS was introduced in 2019. Therefore, our study collects data only on SC, ST, OBC, and general categories.
8 Country Classification as per new criteria (July 1, 2020) https://blogs. worldbank.org/opendata/new-world-bank-country-classifications-income-level-2020-2021, accessed on March 2, 2021.
9 As per census 2011 definition of "kachcha" and "pucka" house.

References

Ahmed, R., Robinson, R., Elsony, A., Thompson, R., Bertel Squire, S., Malmborg, R., Burney, P., & Mortimer, K. (2018). A comparison of smartphone and paper data-collection tools in the Burden of Obstructive Lung Disease (BOLD) study in Gezira state, Sudan. *PloS One, 13*(3), 1–15.

Ahuja, M. (2016). Age of menopause and determinants of menopause age: A PAN India survey by IMS. *Journal of Midlife Health, 7*(3), 126–131. https://doi.org/10.4103/0976-7800.191012

Bhatia, J. C., & Cleland, J. (1995). Self-Reported symptoms of gynecological morbidity and their treatment in South India. *Studies in Family Planning, 26*(4), 203–216.

Brander, M. S. (1942). Tampons in menstruation. *The British Medical Journal, 1*(4239), 452–452. JSTOR.

Browne, R. V., Ellis, N. C., & Hird, J. A. (1989). The pre-menstrual syndrome: A study in people with a mental handicap. *Psychiatric Bulletin, 13*, 363–365.

Bulsari, S., Siddiqui, N., Saeed, S., & Sarfraz, H. (2020). *Adapting to the new normal in survey research. Covid-19, Education and Educational Research.* https://www.bera.ac.uk/blog/adapting-to-the-new-normal-in-survey-research

Cardwell, M. G. (1942). Tampons in menstruation. *The British Medical Journal, 1*(4242), 537.

Chaghlana, S. P. S., Amaravadhi, S. R., Mazodi, S. D., & Lokusu, V. S. (2019). Determinants of menstrual hygiene among school-going adolescent girls in urban areas of Hyderabad. *International Journal of Community Medicine and Public Health, 6*(5), 2211–2215.

Chenery, H. B., & Taylor, L. (1968). Development patterns: Among countries and over time. *The Review of Economics and Statistics, 50*(4), 391–416. JSTOR. https://doi.org/10.2307/1926806

Coleman, J. S. (1958). Relational analysis: The study of social organizations with survey methods. *Human Organization, 17*(4), 28–36. JSTOR.

Couper, M. P. (2000). Review: Web surveys: A review of issues and approaches. *The Public Opinion Quarterly, 64*(4), 464–494. JSTOR.

De Leeuw, E. D. (2012). Counting and measuring online: The quality of internet surveys. *BMS: Bulletin of Sociological Methodology/Bulletin de Méthodologie Sociologique, 114*, 68–78. JSTOR.

Durr-e-Nayab. (2005). Reproductive tract infections among women in Pakistan: An urban case study. *The Pakistan Development Review, 44*(2), 131–158. JSTOR.

Frippiat, D., Marquis, N., & Wiles-Portier, E. (2010). Web surveys in the social sciences: An overview. *Population (English Edition, 2002–), 65*(2), 285–311.

Ganie, M. A., Vasudevan, V., Wani, I. A., Baba, M. S., Arif, T., & Rashid, A. (2019). Epidemiology, pathogenesis, genetics & management of polycystic ovary syndrome in India. *Indian Journal of Medical Research, 150*(4), 333–344. https://doi.org/10.4103/ijmr.IJMR_1937_17

Garg, S., Sharma, N., & Sahay, R. (2001). Socio-cultural aspects of menstruation in an urban slum in Delhi, India. *Reproductive Health Matters, 9*(17), 16–25.

Goodman, L. A. (1961). Snowball sampling. *The Annals of Mathematical Statistics, 32*(1), 148–170. JSTOR.

Government of Gujarat. (2020). *Socioeconomic review 2019–20, Gujarat state* (Budget Publication No. 34). Directorate of Economics and Statistics.

Granovetter, M. (1976). Network sampling: Some first steps. *American Journal of Sociology, 81*(6), 1287–1303. JSTOR.

Handcock, M. S., & Gile, K. J. (2011). Comment: On the concept of snowball sampling. *Sociological Methodology, 41*, 367–371. JSTOR.

Henderson, V. (2003). The urbanization process and economic growth: The so-what question. *Journal of Economic Growth, 8*(1), 47–71.

ILO. (2020). *COVID-19 Impact on the collection of labour market statistics*. ILOSTAT. https://ilostat.ilo.org/topics/covid-19/covid-19-impact-on-labour-market-statistics/

Lucas, J. W. (2003). Theory-testing, generalization, and the problem of external validity. *Sociological Theory, 21*(3), 236–253. JSTOR.

Lucas, Jr, R. E. (1988). On the mechanics of economic development. *Journal of Monetary Economics, 22*, 3–42.

Marcano Belisario, J. S., Jamesek, J., Huckvale, K., & Car, J. (2015). Comparison of self-administered survey questionnaire responses collected using mobile apps versus other methods. *Cochrane Database of Systematic Reviews (Online)*. https://www.researchgate.net/publication/280497078_Comparison_of_self-administered_survey_questionnaire_responses_collected_using_mobile_apps_versus_other_methods?enrichId=rgreq-69b76c775e74f6f4e14b3ac9e03b064d-XXX&enrichSource=Y292ZXJQYWdlOzI4MDQ5NzA3ODtBUzo1OTE4NDM1MDcwNDg0NDlAMTUxODExNzg5Njg1OA%3D%3D&el=1_x_2&_esc=publicationCoverPdf

MoHFW, & IIPS. (2020). *Fact sheets: Key indicators: 22 States/UT from phase—I: 2019–2020* (NFHS–5). Indian Institute of Public Health.

Mulgaonkar, V. B. (1996). Reproductive health of women in urban slums of Bombay. *Social Change, 26*(3 & 4), 137–156.

NIPCCD. (2014). *Improvement in knowledge and practices of adolescent girls regarding reproductive health with special emphasis on hygiene during menstruation in five years*. National Institute of Public Cooperation and Child Development.

Patel, H. R., & Patel, R. R. (2016). A cross sectional study on menstruation and menstrual hygiene among medical students of Valsad, Gujarat. *International Journal of Reproduction, Contraception, Obstetrics and Gynecology, 5*(12), 4297–4302.

Pathak, P. K., Tripathi, N., & Subramanian, S. V. (2014). Secular trends in Menarcheal Age in India: Evidence from the Indian Human Development Survey. *PloS One, 9*(11), 1–13. https://doi.org/10.1371/journal.pone.0111027

Period Products (Free Provision) (Scotland) Act, Scotland Parliament (2021).

Press Information Bureau. (2018). GST council recommends GST rates reduction on several goods and for specified handicraft. *Governemtn of India*. https://pib.gov.in/Pressreleaseshare.aspx?PRID=1539574

Romer, P. M. (1986). Increasing returns and long-run growth. *Journal of Political Economy, 94*(5), 1002–1037. JSTOR.

Santra, S. (2017). Assessment of knowledge regarding menstruation and practices related to maintenance of menstrual hygiene among the women of reproductive age group in a slum of Kolkata, West Bengal, India. *International Journal of Community Medicine and Public Health, 4*(3), 708–712. JSTOR.

Sengupta, A., Bhattacharya, P., & Guha, S. (2020). More women opting for safe menstrual practices, NFHS-5 data shows. *Down to Earth, December 23*. https://www.downtoearth.org.in/blog/health/more-women-opting-for-safe-menstrual-practices-nfhs-5-data-shows-74758

Sridhar, N. (2019). *Menstruation across cultures: A historical perspective.* Vitasta Publishing Pvt Ltd.

The Essential Commodities (Amendment) Bill, 36 of 2022, Parliament, 2 (2022). https://sansad.in/getFile/BillsTexts/LSBillTexts/Asintroduced/ 36%20of%202022%20as%20introduced.pdf?source=legislation#:~: text=(i)%20sanitary%20pads%3B%20(,iii)%20menstrual%20 cups%3B.%E2%80%9D&text=%E2%80%9C(9)%20menstrual%20 hygiene%20products.

UNFPA. (2022). Menstrual hygiene among adolescent girls: Key insights from the NFHS-5 (2019–21). *Analytical Paper Series, 2.* https://india. unfpa.org/sites/default/files/pub-pdf/analytical_series_2_-_menstrual_ hygiene_among_adolescents_-_insights_from_nfhs-5_final.docx.pdf

Velayutham, K., Selvan, S. S. A., & Unnikrishnan, A. (2015). Prevalence of thyroid dysfunction among young females in a South Indian population. *Indian Journal of Endocrinology and Metabolism, 19*(6), 781–784.

Verma, R. (2019). Modi declares India ODF: Sustaining the status the next big challenge. *Down to Earth*, October 3.

WHO. (2022). *WHO Statement on Menstrual Health and Rights: 50th Session of the Human Rights Council Panel Discussion on Menstrual Hygiene Management, Human Rights and Gender Equality.* https://www.who.int/ news/item/22-06-2022-who-statement-on-menstrual-health-and-rights

Winkler, I. T. (2020). Introduction: Menstruation as fundamental. In C. Bobel, I. T. Winkler, B. Fahs, K. A. Hasson, E. A. Kissling, & T.-A. Roberts (Eds.), *The Palgrave handbook of critical menstruation studies* (pp. 9–13). Springer Singapore. https://doi.org/10.1007/978-981-15-0614-7_2

2 Awareness, Beliefs, and Taboos around Menstruation

There are many beliefs and taboos associated with menstruation. However, if one looks through the history, one may find reasons for having those taboos in place, which are apparently not relevant in the present times. Historically, monthly menstrual cycles were referred to using terms like "the flowers", "the terms or periods", "the course", "the months". Medics used the terms "menstris", "menses", "catamenia", "monthly evacuations" or "natural purgation", way back, in the seventeenth century (Crawford, 1981). All these terms subtly meant the physical weakness in women, the sickness experienced by women, a monthly disease, or a monthly infirmity. Naturally, infirmity emanated from the fact that absorbents with high blood-absorbing capacity were not available then, and therefore, women either used old clothes, or natural material like mud, sand, leaves, etc., or nothing at all to soak the menstrual blood. They argue that it is not possible to attribute one single reason that determined the position of women in the society, but men's attitudes to menstruation played an important role in shaping their attitudes towards women. It was immaterial whether menstruation implied women's inferiority to men, or vice versa, menstruation did confine the role of women to domestic duties only. Beliefs about the inferiority of women shaped ideas about menstruation, and beliefs about menstruation reinforced that women are inferior to men. On the other hand, Apffel-Marglin (2020) argues that when little was known about menstruation, it was said that women had some magic powers, which caused menstruation and childbirth, and men were perhaps jealous of them. Beliefs about menstruation were largely based on experiences, which in turn, would have become a cultural norm. However, with the passage of time, this would have become a taboo or a superstition. They further argue that while menstruation is taught as a biological phenomenon in today's world, it is equally important to examine it from the anthropological perspective to understand it in its totality.

NIPCCD (2014) has succinctly summarised the menstruation taboos across religions: they quote "The Old Testament of The Bible" that a menstruating woman is impure and that most things she touches become

DOI: 10.4324/9781003473961-2

unclean. However, today, these menstruation taboos have largely become obsolete, except for Russians and Coptic Christians of Ethiopia. There were similar taboos in Judaism too but are now practiced only by religious people. Sridhar (2019) discusses in detail, the anthropological interpretations of menstrual cycles across cultures and religions. They observe that Hinduism and Jainism continue to practice seclusion and restrictions on consumption of certain foods. They further observe that Islam has its own set of taboos on religious activities and fasting, but women are not required to observe seclusion. Buddhism is spread across India, China, Tibet, Japan, and Southeast Asia. There are cultural differences, however little, across these countries, but the women are considered impure and refrain from religious activities and praying or meditating during their menstrual cycles. Not only across religions, but women are required to observe seclusion, and the menstruation is associated with impurity across cultures. This includes different tribes of North America, Canada, Central and South America, Africa, Australia, South Asia, South-East Asia, and Europe. The only exception among this is Sikhism, where no restrictions are followed, and a menstruating woman is permitted even to the Gurudwara (their place of worship) and to perform the religious activities.

Adherence to beliefs and taboos can have an impact on hygiene practices, as well as on physical and mental health. Therefore, awareness about menstruation helps in understanding the importance of menstruation and hygiene practices associated with it. This is because certain taboos, like staying in a secluded place during menstruation result in unhygienic use or disposal of absorbents (Krishnan and Twigg, 2016). Some hygienic practices, like using a disinfectant to clean private parts, changing a sanitary napkin/absorbent every four hours, etc., involve costs. On the other hand, beliefs like staying away from work/school could have long-term implications in terms of performance and career selection/progression. Taboos that hinder hygienic practices also result in infections, that can have far-reaching consequences. Therefore, it is required to study the beliefs and taboos still prevalent in the society. Also, examining the extent of awareness about menstruation and menstrual hygiene is helpful to work out interventions and policy recommendations about spreading/increasing awareness about menstruation, and menstrual health and hygiene.

In this chapter, we examine the patterns in awareness, beliefs, and taboos around menstruation from the published literature, augmented with the empirical evidence that emerges from our study. We haven't attempted either of systematic, umbrella, or rapid reviews for this study on the patterns in awareness. This is because the primary objective of our study is to work out a cost estimate to inform the policy on interventions in women's reproductive health/menstrual health. Therefore, our search attempts to identify the set of menstrual awareness parameters, which we use in our study to develop the module on awareness, beliefs, and taboos for the survey instrument (questionnaire). Our search was performed largely on JSTOR in 2019 and later updated in 2024 with additional searching in Google Scholar, to include the latest literature. We have not made use of grey literature. Search terms were "menstruation awareness", "menstruation beliefs and taboos", and "menstruation awareness, beliefs and taboos". We have tried to include all articles until saturation.[1]

We included articles that discussed awareness at an individual level. Moreover, we examine awareness specifically of menstruation, menstrual hygiene, or its management, not on awareness about any other things, for example, oral health during menstrual cycles. We have been open in terms of selecting articles on beliefs and taboos, both at the individual and community level.

This chapter is organised in three broad sections: Section 2.1 discusses the extent to which girls/women are aware of menstruation and menstrual hygiene. Section 2.2 discusses the beliefs and taboos associated with menstruation. We discuss the findings from our study in the context of existing literature on awareness, and beliefs and taboos, in respective sections. We have also attempted to examine whether each of awareness, beliefs and taboos differ across age, education, and socioeconomic status. Section 2.3 presents the key findings on awareness, beliefs, and taboos based on our research.

2.1 Awareness around Menstruation and Menstrual Hygiene Practices

A girl's body experiences physical and hormonal changes around menarche. This gives an indication of the girl entering her puberty. She experiences hair growth in the pubic area and armpits. Her breasts start developing before she gets her menarche (Lacroix et al., 2021; Robeva and Kumanov, 2016). We review the existing literature in this section to discern patterns in awareness about the absorbent for soaking menstrual blood, frequency of change of absorbent material, disposal of menstrual products, washing hands after using the washroom, as well as washing the private parts during menstruation, largely among adolescent girls. Most published literature is undertaken in the Indian context. However, we have also included the studies undertaken in the context of other Asian and developing countries.

We also use these patterns to seek insights for designing the questions for our survey, to gauge awareness around menstruation among women of reproductive age. The questions of our survey are also informed by the patient and public involvement (PPI)[2] and professional experts' (doctors'/gynaecologists') inputs. The interview guide for gynaecologists and doctors practicing reproductive health is provided in Appendix A. We also discuss the results of our study on awareness around menstruation by women across socioeconomic and demographic groups.

2.1.1 Patterns in Awareness around Menstruation based on Available Literature

There is mixed evidence on women's awareness about menstruation in totality. Most of the studies reveal that girls use disposable sanitary napkins and follow hygienic cleaning practices. The only exception was Kaur et al. (2018), where most girls in rural areas used reusable cloth as an absorbent. Moreover, in Bangladesh, it was found that only one-fourth of the women make use of a specialised absorbent like sanitary napkins (Afiaz and Biswas, 2021). On the other hand, Al Mutairi and Jahan (2021) observed that most girls were not having satisfactory hygienic practices; Michael et al. (2020) found that a majority of girls in Pakistan did not take baths, or clean their private parts.

Kotian et al. (2017) observe that in Mangalore, Karnataka, women in the age group of 15–24 were better informed about menstruation and menarche in comparison with those who were in the age group of 25–35 years. Anusree et al. (2014); Chadalawala and Kala (2016); Patavegar et al. (2014); Sarkar et al. (2017); and Sreenivasa et al. (2017) also find that a vast majority of their survey respondents, who were in the age group of 13–24, were aware of menstruation prior to their menarche. These Indian studies were based in Mangalore, rural Vijayawada, New Delhi, Chinsurah-Mogra block of Hoogli district of West Bengal, and urban Bangalore, respectively. Tribal girls aged 13–18 years in Khunti district of Jharkhand, India, and girls in the age group of 10–19 years in Chitwan, Nepal, were also found to have required awareness about menstruation prior to menarche (Kumari et al., 2021; Neupane et al., 2020). The girls in Nepal, Neupane et al. (2020) were also aware of what causes menstrual cycles. There are similar observations found in Pakistan (Michael et al., 2020), Saudi Arabia (Al Mutairi and Jahan, 2021), and Central Ethiopia (Bulto, 2021). Contrary to these, the studies undertaken by Arshad Ali et al. (2020); Mahajan and Kaushal (2017); Prajapati and Patel (2015); and Thakur et al. (2014) found that very few young girls in the age group of 15–24 years had information about menstruation before their menarche. These studies were undertaken in Pakistan and in Shimla, Himachal Pradesh, Gandinagar, Gujarat and lower socioeconomic area of Naigaon, Mumbai in India.

Additionally, it was observed that despite widespread awareness about the process of menstruation, in terms of age of menarche, duration and frequency of menstrual cycles, and management of menstrual cycles, majority of girls did not know the source of menstrual blood, with the exceptions of Chadalawala and Kala (2016) and Michael et al. (2020).

Girls from rural Malaysia also expressed that they needed more information on menstruation and menstrual hygiene, though they were found to use (and regularly change) the sanitary napkins and also dispose off, the used ones, appropriately (Abdul Hakim et al., 2024).

Despite the mixed evidence on awareness about menstruation, mother has unequivocally been the source of information about menstruation and menarche. However, Kotian et al. (2017); Mahajan and Kaushal (2017); and Patavegar et al. (2014) provide additional insights in terms of mother's education level, and mother's occupation (Anusree et al., 2014), contributing to the dissemination of information and extent of awareness about menstruation and menarche. Contrary to these studies, Prajapati and Patel (2015) found no association between mother's education and girls' awareness around menstruation.

It may be noted that most studies have not examined the difference in extent of awareness around menstruation by socioeconomic profile of the respondents, except that Arumugam et al. (2014) explored the rural-urban differences in awareness about menstruation and found that the awareness around menstruation was higher in rural areas by around 20 percentage points, in comparison with the urban areas. On the other hand, Kumar et al. (2024) observe that education, residential area, or age has no influence on awareness regarding menstruation.

2.1.2 Patterns in Awareness around Menstruation based on our Survey Results

We examined the pattern of awareness among women of reproductive age groups as well as the association of different socioeconomic factors on awareness surrounding menstruation. We also explored the source of awareness about menstruation and discussed it in this section. Figure 2.1 shows the patterns in these women's source of menstruation awareness prior to menarche, and who guided the women during the first cycle and the subsequent yet initial cycles of menstruation.

It can be seen from Figure 2.1 that 80.49 per cent girls in the first menstrual cycle and 83.32 per cent girls in the initial cycles received guidance from their mothers. All these girls had information about menstruation prior to menarche.

However, lesser proportion of women were found to have awareness about the source of menstrual blood and about the number of days to be considered normal between two menstrual cycles. Figure 2.2 further shows that there is relatively lower awareness about the source of menstrual blood; only 47.61 per cent respondents have the correct information about it. Also, 55.9 per cent women are aware about the duration between two consecutive menstrual cycles (normally, between 26 days and 32 days).

Some physical changes in girls, like growth of pubic hair, breasts development, etc., take place before and just after menarche (Lacroix et al., 2021; Robeva and Kumanov, 2016). Respondents were asked whether they are aware about these changes that take place at puberty. A large proportion of women were found to have awareness about the physical

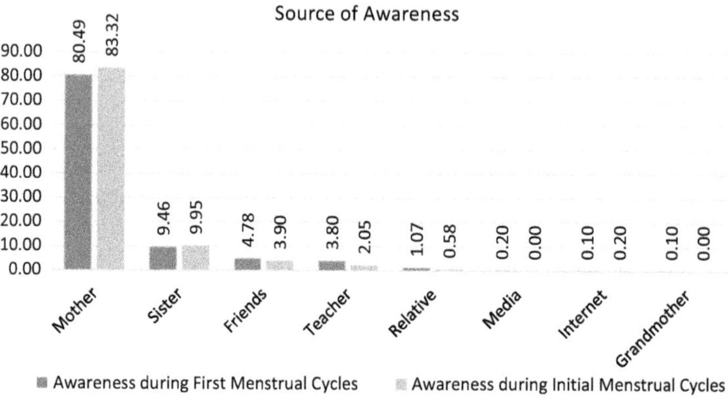

Figure 2.1 Source of Awareness about Menstruation during the First and Initial Menstrual Cycles.

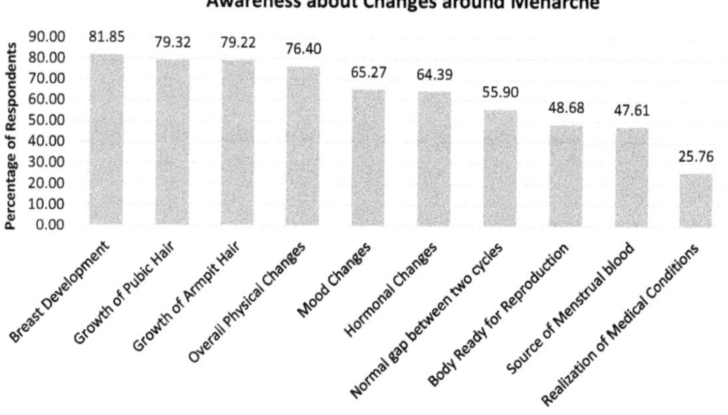

Figure 2.2 Awareness about Physical Changes in the Body, Hormonal Changes, Realisation of Medical Condition during Menarche and Menstruation.

changes in the body around menarche. Out of 1025 respondents, more than 65 per cent women were aware about the physical and hormonal changes. However, less than 50 per cent of women were aware that their bodies get ready for reproduction after menarche.

Literature suggests that some medical conditions like endometriosis (Engemise et al., 2010; Darrow et al., 1993) and PCOD (Brook et al., 1988) are diagnosed soon after menarche. Figure 2.2 shows that relatively fewer respondents (25.76 per cent) had awareness about medical conditions at menarche. It is quite likely that many of them might have no known medical conditions so far; only 25.76 per cent were aware that they can come to know about medical conditions, if any, on attaining puberty.

This study also examines the socioeconomic factors that influence awareness. In order to examine the impact of socioeconomic factors on menstruation awareness, awareness score is developed. Awareness score is generated by summing up the number of "Yes" responses for each aspect of awareness (10 items presented in above Figure). Thus, awareness score ranged between 10 and 0. Summary statistics of awareness score are given in Table 2.1:

Awareness score is regressed against the socioeconomic factors. We have used Forward Stepwise Regression[3] to choose the relevant socioeconomic factors for all the models in this chapter. We have found the optimal model with six out of eight socioeconomic factors. Awareness score is then regressed against these six relevant socioeconomic factors. The results of regression are given in Table 2.2:

Table 2.1 Awareness Score Statistics

Statistics	*Awareness Score*
Min	0
1st Quartile	5
Median	7
Mean	6.244
3rd Quartile	8
Max.	10

Table 2.2 Influence of Socioeconomic Factors on Menstruation Awareness Scores

*Socioeconomic Factors**	*Estimate*	*Std. Error*	*t-value*	*Pr(>|t|)*
(Intercept)	7.1066	0.5460	13.017	< 0.0001
1. Occupation: Working (indoors)	0.5276	0.3696	−1.428	0.1537
Other work (outdoors)	0.7318	0.3791	1.93	0.0538
College-going student	0.3176	0.3889	0.817	0.4144
Non-working	−1.5409	0.3813	−4.041	<0.0001
School-going student	−0.5426	0.4837	−1.122	0.2623
2. Region: Urban	0.9992	0.1614	6.191	< 0.0001
3. Development status: Developing	−0.8304	0.1877	−4.424	< 0.0001
Tribal	−0.3749	0.2403	−1.56	0.1191
Undefined	−0.9971	0.6226	−1.602	0.1096

(*Continued*)

Table 2.2 (Continued)

Socioeconomic Factors*	Estimate	Std. Error	t-value	Pr(>|t|)
4. Family type: Joint	−0.8028	0.4028	−1.993	0.0465
Nuclear	−0.2569	0.4054	−0.634	0.5264
5. Education: Illiterate	−1.6343	0.6129	−2.667	0.0078
Literate (with or without schooling)	−1.2934	0.4297	−3.01	0.0027
Up to 12th grade	−0.4219	0.23	−1.835	0.0668.
Up to primary	−0.4102	0.3489	−1.176	0.2401
Up to secondary	−0.6907	0.2682	−2.575	0.0102
6. Marital Status: Ever married (Married but currently not with the husband)	−0.5101	0.4727	−1.079	0.2808
Never married	−0.343	0.1891	−1.814	0.0700

Residual standard error: 2.261 on 1006 degrees of freedom
Multiple R-squared: 0.2545, Adjusted R-squared: 0.2411
F-statistic: 19.08 on 18 and 1006 df, p-value: < 0.0001

* For categorical predictors like education, occupation, region, development status, etc., the coefficients are interpreted in the context of a reference category. In the case of categorical predictors, the coefficients will be calculated for all the categories, except the reference category. In Table 2.2, the reference category for occupation is "Working outdoor for manual labour". The reference category for region is "rural", that for development status is "developed", for family type, it is "extended". The reference category for education is "higher education", and for marital status it is "married". These reference categories remain the same across all the regressions used in this chapter. In case of additional predictors, the reference categories are given in footnotes of the respective regressions.

The F-statistic and the associated p-value in Table 2.2 show that the model is a good fit and statistically significant. While the adjusted R-square is low (0.2545), the objective of this regression is to examine the influence of different socioeconomic factors.

The results show that non-working women have a lower awareness score by 1.54 in comparison with women working outdoors for manual labour (reference category). The awareness score of women across all other occupations is the same as that for women working outdoors for manual labour.

The results further show that urban women have clearly higher awareness about menstruation and menstrual hygiene compared to rural women, as can be seen from the positive and statistically significant coefficient of the urban region. Also, the negative and statistically significant coefficients for the developing districts imply lower awareness in developing districts in comparison with developed. Though a non-significant coefficient for tribal districts shows no difference in awareness between developed and tribal districts.

A statistically significant, negative coefficient for non-working women shows a clearly low awareness compared to women in the reference category (working outdoors for manual labour). Women in all other categories of occupation do not have a statistically significant difference in awareness compared to women working outdoors for manual labour.

Table 2.2 shows that education has an influence on awareness around menstruation. Statistically significant negative coefficients for illiterate, just literate (with or without schooling), and secondary education show that these women are less aware compared to those who have higher education attainment. However, there is no statistically significant difference in awareness for women with primary education and those with higher education.

In comparison with women from extended families, the ones in joint families have lower awareness, but this difference is not statistically significant for women in nuclear families.

There is no difference in awareness across marital status categories (Ever vs. Never married).

Thus, non-working women, women from rural areas, developing districts, and those who live in joint families have a relatively lower level of awareness. By and large, there is an increase in awareness with the level of education. Marital status has no influence on awareness around menstruation.

2.2 Beliefs and Taboos

Our study also examines a wide range of beliefs and taboos associated with menstruation. People hold beliefs like menstruation is impious, absence of menstruation indicates pregnancy, pain during menstruation indicates sickness, etc. The beliefs and taboos associated with menstruation result in poor awareness about menstruation, which in turn, results in poor menstruation management and menstrual hygiene. Itching, infections, and in extreme cases, internal swelling or even abscess could result from poor menstrual hygiene. Therefore, understanding the nature of beliefs, and the sets of taboos is important to identify the factors that might adversely impact menstrual hygiene. The influence of socioeconomic background on beliefs and taboos score is also examined and discussed in this section.

2.2.1 Patterns in Beliefs and Taboos based on Existing Literature

There is a steep rise in published literature on beliefs and taboos around menstruation in the recent past. However, until 2019, the time when the questionnaire for this study was designed, there were relatively fewer studies in this area. So, we began with one of the oldest studies (Montgomery, 1974), which examined various dimensions of menstrual practices using an ethnographic approach to understand each dimension in detail. While we do not plan to segregate the literature available until 2019, and after, we do highlight the additional information on beliefs and taboos obtained from the relatively newer studies.

Montgomery (1974) refers to Ashley-Montagu (1940) and Stephens (1961), and underpins the psychological and sociological reasons for the emergence of menstrual beliefs and taboos. This implies that scientific discussion on menstrual beliefs and taboos dates back to at least 1940.

We do not attempt to go into the foundations of the evolution of menstrual taboos since the objective of our study is to understand the effects of existing taboos on menstrual hygiene practices, which in turn have a bearing on costs. Montgomery's ethnographic study covers 44 different societies to examine various beliefs resulting in taboos for women during menstruation; these are listed as follows:

- Menstrual fluid is unpleasant, contaminating/dangerous.
- Menstruating women cannot have sexual intercourse.
- Menstruating women should restrict their movements.
- Personal belongings of men, weapons, instruments used in agriculture and fishing, craft tools, certain crops termed as "men's crops", religious emblems, and shrines should not be touched by menstruating women.
- Hams or flowers get spoilt if menstruating women handle them.
- Cooking for men is prohibited during menstruation.
- Menstruating women should be confined to separate huts (called, the menstrual huts) during their menstrual cycles.

Additionally, their study also found that medical conditions like epilepsy and migraine manifest/get identified at menarche, and medical conditions such as multiple sclerosis, myasthenia gravis, angiovascular disorders, rashes, and urticaria manifest at menarche and worsen during the menstrual cycles. These details have helped us to understand and build a conceptual framework for cost components of menstrual hygiene (discussed in Chapters 3 and 4), the core objective of our study. However, in this section, we focus on discussing beliefs and taboos around menstruation.

As early as 1850s, when the branch of gynaecology was given recognition, all the issues associated with pelvic organs were collectively described as women's diseases, and menstruation was perceived even by physicians as "unfortunate, unpleasant and distasteful subject to address" and that even women should be spared from this curse (Strange, 2000). We did examine in our study, whether women consider menstruation, absence/irregularities in menstruation, and pain during menstruation as a disease or not.

The commonest taboo associated with menstruation is refraining from religious activities and rituals, entering the areas designated for worship, going to temple, or seeing God's images. This was found across societies and socioeconomic status in Asian countries.

Entering the kitchen, cooking, or overall kitchen chores (Das, 2008; Dündar and Aksu, 2022; Khan et al., 2024; Mahon and Fernandes, 2010; Srinivasa and Manasa, 2017; Uppal et al., 2022) and touching food, especially pickles (Kumar and Srivastava, 2011) were common taboos across the communities and cultures. There are beliefs about the type of foods that a woman can consume during her menstrual cycles. This belief has resulted in a taboo around consuming certain types of food (Al Mutairi and Jahan, 2021; Arshad Ali et al., 2020; Garg et al., 2001; Patavegar et al., 2014; Syed Abdullah, 2022; Uppal et al., 2022; van Eijk et al., 2016), and that some foods were consumed to prepone or postpone the menstrual cycles (Arumugam et al., 2014). Women are refrained from doing household chores other than cooking (Mahon and Fernandes, 2010; van Eijk et al., 2016).

Women are also restrained from undertaking physical activities (Deo and Ghattargi, 2005), playing/exercising (Mahon and Fernandes, 2010; van Eijk et al., 2016), and travelling or travelling alone (Deo and Ghattargi, 2005) during their menstrual cycles.

All these taboos – refraining from kitchen and household chores, physical activities, and consuming certain types of foods, over and above taboos on religious activities – were observed in Sreenivasa et al. (2017) and (L. Ganguly et al., 2021). Mukherjee et al. (2020) found that most Hindu girls in Nepal, apart from refraining themselves from praying, practiced purifying kitchen, bed, bedsheets, and other household things on the fourth day of their menstrual cycles. They also observed that girls from Janajati (tribal/indigenous castes) and those with higher education (Masters) were less likely to observe these taboos. Ganguly et al. (2024) found that education, family environment, cultural background, and beliefs influence taboos surrounding menstruation. Contrary to this, (Kumar et al., 2024) found no association between education and taboos.

Other taboos are in smaller proportions and they include abstaining from sex or sleeping with husband, touching boys/men or even interacting with them (Deo and Ghattargi, 2005; Dündar and Aksu, 2022; Garg et al., 2001; Khan et al., 2024; Mahon and Fernandes, 2010).

Garg et al. (2001) further observed that the restrictions on bathing were observed by women in certain religions. Some respondents are also found to associate witchcraft with menstrual cloth; they had to bury the cloth in the pit latrine; otherwise, witchcraft would make them infertile and they would be deprived of giving birth to babies.

Deo and Ghattargi (2005) also found that the girls observed restrictions on the type of clothing; though there is no detailed discussion on the nature of clothing to be worn/avoided during the menstrual cycles. Arumugam et al. (2014) found that there are taboos on going to the school and touching flowers.

However, Deo and Ghattargi (2005) additionally found that taboos are more prevalent in rural areas as compared to urban areas. The findings of Arumugam et al. (2014) also show that these taboos are fast fading away in urban areas.

Uppal et al. (2022) found that menstruation was believed to be dirty or dangerous. In some societies, there is a stigma around women who do not menstruate and are often abused and cursed. There are taboos on touching or cleaning private parts during menstrual cycles in certain cultures. Women are believed to be ready for marriage and having sex soon after attaining menarche. Also, women are believed to be physically and emotionally weak during their menstrual cycles.

Syed Abdullah (2022) found in her study that the practice of seclusion during menstrual cycles still persists in Malaysia. Our study also explores the practice of seclusion, in addition to other beliefs and taboos discussed above.

Dündar and Aksu (2022) also found that menstruating women abstain from visiting the newborn babies, whereas Khan et al. (2024) found that they are not allowed to attend funerals.

It may be noted that most studies are undertaken in different states/ regions of India, and a few of them are from other Asian countries. Also, in comparison with India, fewer studies are undertaken in other Asian countries.

2.2.2 *Beliefs around Menstruation*

Based on the above literature published until 2019 and PPI work, eight different beliefs associated with menstruation are identified, and the respondents were asked to rate on a five-point Likert-type scale, which ranged from strongly believe to strongly disbelieve. These are:

- Menstruation is a disease.
- Menstruation is impious.
- Absence of menstrual cycles is pregnancy.
- Menstrual blood contains dangerous substances.
- Pain during menstruation indicates sickness.
- Irregularity in menstrual cycles is a disease.
- Regularity in menstrual cycles but shorter than 28 days is a disease.
- Regularity in menstrual cycles but longer than 28 days is a disease.

Figures 2.3 and 2.4 show the distribution of beliefs around menstruation and menstrual cycles.

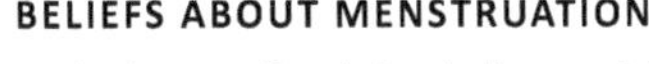

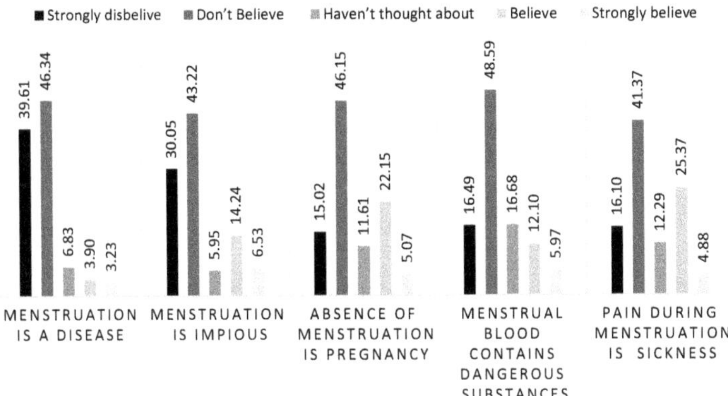

Figure 2.3 Beliefs about Menstruation.

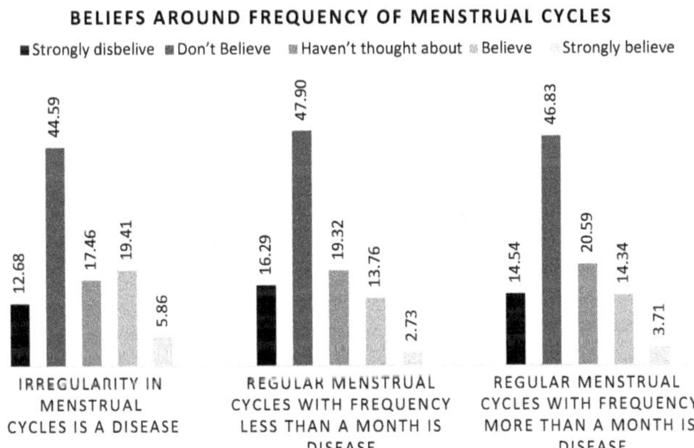

Figure 2.4 Beliefs about Menstrual Cycles.

Figure 2.3 shows that more than 70 per cent women do not consider menstruation to be a disease or impious. More than 50 per cent do not consider menstruation as sickness or that menstrual blood contains dangerous substances. However, 60 per cent or more believe that the absence of menstrual cycles means pregnancy. Figure 2.4 shows that out of 1025 respondents, more than 50 per cent do not believe irregularities in menstrual cycles or a gap of more (or less) than a month between two cycles to be a disease. This means that very few women hold these beliefs about menstruation.

Table 2.3 Beliefs Score Statistics

Statistics	Belief Score
Min	−0.6587
1st quartile	−0.0962
Median	0.0288
Mean	0
3rd quartile	0.0913
Max.	0.3413

The Beliefs model examines the impact of socioeconomic background on beliefs associated with menstruation.

Belief score is developed by first, assigning a value of −2 to strongly believe, −1 to believe, 0 to haven't thought about it, 1 to don't believe, and 2 to strongly disbelieve. There are in total eight belief measures. We calculated the individual z-scores across these eight belief measures, and a mean of these eight individual z-scores to get the belief score. Belief score statistics are given in Table 2.3.

Thus, a higher belief score indicates milder beliefs, and a lower score indicates stronger beliefs. Belief score is regressed against the factors defining the socioeconomic background of the respondents. We regress belief scores on socioeconomic factors to see which ones are associated with beliefs around menstruation and menstrual cycles. Before moving on to the regression results in Table 2.4, we would like to explain that we have defined Age, Occupation, Region, Development Status, Family Type, Education Level, Marital Status, and Religion as predictors to all the regression models presented in this chapter. Forward selection algorithm used to undertake regression would include one predictor at each stage and would stop when adding a predictor stops contributing to the model. This is done in order to ensure the principle of parsimony (Gori et al., 2024; Gujarati et al., 2011). However, it may also be noted that Age has not been included in any of the models by forward regression algorithm. This implies that the addition of age to any of these models does not make any difference in the explanatory power of the model.

The results of the regression are given in Table 2.4.

Forward stepwise regression results show that four out of eight socioeconomic factors influence beliefs around menstruation.

All the coefficients of education are negative, implying higher belief scores for higher education. Higher belief scores indicate milder beliefs. Thus, education helps in having a better understanding of the menstruation process and thus frees one from the beliefs associated with it.

In comparison with women working outdoors for manual labour, the belief score for non-working women is higher. The belief score of all working women (indoors or outdoors) is higher compared to that of women

Table 2.4 Impact of Socioeconomic Factors on Beliefs associated with Menstruation

Socioeconomic Factors*	Estimate	Std. Error	t-value	Pr(>|t|)
(Intercept)	−0.0523	0.0321	−1.629	0.1037
1. Education: Illiterate	−0.2857	0.0517	−5.522	<0.0001
Education: Literate (with or without schooling)	−0.0039	0.0361	−0.108	
Up to 12th grade	−0.0279	0.0193	−1.446	0.1485
Up to primary	−0.1021	0.0293	−3.481	0.0005
Up to secondary	−0.0693	0.0224	−3.098	0.0020
2. Occupation: Working (indoors)	0.0580	0.0311	1.864	0.0627
Other work (outdoors)	0.0601	0.0320	1.88	0.0604
College-going student	−0.0233	0.0319	−0.731	0.4651
Non-working	0.0890	0.0321	2.772	0.0057
School-going student	0.0040	0.0391	0.102	0.9184
3. Region: Urban	0.0493	0.0136	3.623	0.0003
4. Development status: Developing	0.0539	0.0158	3.41	0.0007
Tribal	−0.0265	0.0203	−1.307	0.1914
Undefined	0.0640	0.0526	1.216	0.2244

Residual standard error: 0.1915 on 1010 degrees of freedom
Multiple R-squared: 0.1387, Adjusted R-squared: 0.1267
F-statistic: 11.62 on 14 and 1010 df, p-value: < 0.0001

* The reference categories for all the variables are same as for the regressions of Table 2.2.

working outdoors for manual labour. This means that in comparison with women working outdoors for manual labour, all other working and non-working women have milder beliefs. The belief scores of school and college-going students are not statistically significant. The underlying reasons to understand this pattern require further exploration.

The belief score for urban women is higher compared to rural women, implying milder beliefs held by rural women in comparison with urban women. This can be explained by the fact that urban women have better exposure and access to information.

The coefficient for tribal districts is not statistically significant, implying no difference between the belief scores of women from developed districts or from tribal districts. However, women in developing districts have a higher belief score, implying milder beliefs as compared to those in developed districts.

In a nutshell, the results reveal that with an increase in education, the beliefs get milder. Women across occupations, except those in manual labour and non-working women, hold milder beliefs. Women from developing districts and living in rural areas have also reported milder beliefs.

2.2.3 Menstruation Taboos

In this study, ten different taboos associated with menstruation are listed, and the respondents were asked to rate them on a three-point scale. – "Can't do", "Can do but I don't", and "Can do and I do". The taboos for which the respondents are asked to rate on the said three-point scale are: attend religious ceremonies, perform religious ceremonies, offer prayers, go to the place of worship, do household work, cook food, touch stored food, live a routine life, sleep on the usual bed, play outdoors/go to work, dance, and do work involving physical strain.

Figure 2.5 shows that more than 50 per cent of respondents refrain from attending religious ceremonies, performing religious ceremonies, and visiting the place of worship, whereas 52 per cent of respondents offer prayers during their menstrual cycles. However, around 25–30 per cent of respondents are of the opinion that there is nothing wrong in attending religious ceremonies, performing religious ceremonies, and visiting the place of worship, but their minds are conditioned since childhood, and hence they refrain from doing so. This indicates that the menstruation taboos are gradually getting milder.

Figure 2.6 shows that more than 75 per cent of respondents do their household work, whereas more than 70 per cent cook during their menstrual cycles, and 53.37 per cent touch stored food. This indicates that the taboos about household work and cooking have also gotten milder.

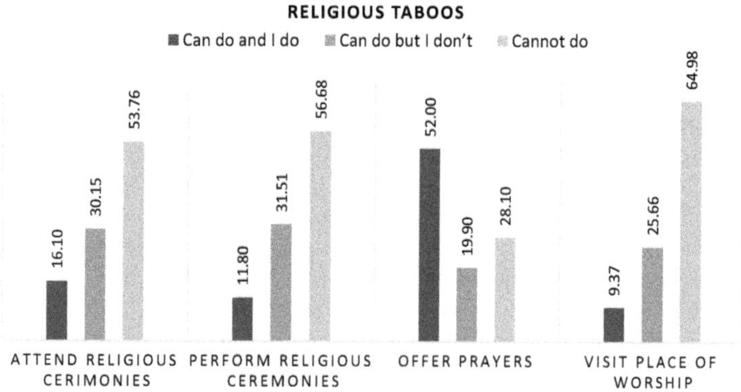

Figure 2.5 Religious Taboos Practiced during Menstrual Cycles.

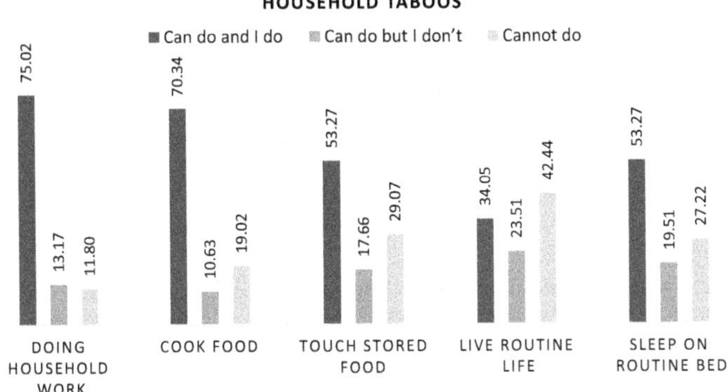

Figure 2.6 Household Taboos Practiced during Menstrual Cycles.

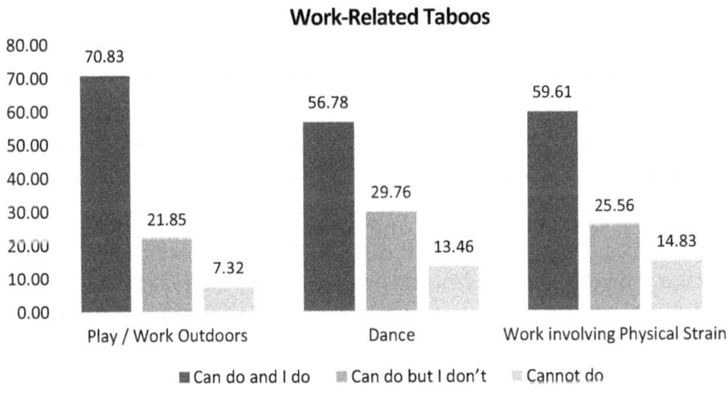

Figure 2.7 Work-Related Taboos Practiced during Menstrual Cycles.

While more than 50 per cent of respondents appear to be sleeping on their routine bed; they do not sleep on a different bed/mattress (normally, kept secluded for menstruating women), a very few seem to be living a routine life. Those not able to live a routine life could be largely, because of the health problems/psychological problems experienced by them during menstruation, and not because of the taboo to live a secluded life. Figure 2.7 shows that more than 70 per cent of respondents seem to be doing outdoor work/playing outdoors, only 56.78 per cent dance and 59.61 per cent do work that involves physical strain.

The response "Can do and I do" shows absence of the taboo. The "Can do but I don't" option indicates a cognitive dissonance between thought and action; thereby, a partial taboo and "Can't do" shows presence of a taboo. Absence of a taboo is coded as 1, partial taboo is coded as 0 and existence of a taboo as −1. Taboo score is developed by taking mean of the z-scores of the responses across these 12 different taboos for each respondent. Higher taboos score indicates milder taboos, and a lower score indicates stricter taboos. Taboo score statistics are given in Table 2.5.

The taboo score is then regressed against the factors defining the socioeconomic background of the respondents. The results of the regression are given in Table 2.6.

Forward selection regression results reveal that religion, region (rural/urban), education, occupation, family type and development status of the districts have bearing on the taboos score. The F-statistic and the associated p-value in Table 2.6 show that the model is statistically significant. Here too, the adjusted R-square is 0.2029 but the objective of this model is only to examine the influence of socioeconomic factors on taboos score.

The coefficients of all categories of religion, except for Sikhs, are negative and statistically significant. This means that women from all religions except Sikhism practice stricter taboos as compared to Christian women. There is no statistically significant difference between the taboo score of Christians and Sikhs.

The coefficient of taboos in urban areas is positive and statistically significant. This indicates that urban women have milder taboos compared to the rural ones.

The coefficients of different categories of education are negative and statistically significant, except for those who are just literate (with or without schooling). This indicates that the intensity of taboos gets stricter with education. This is contrary to the logic, since education is expected to liberate an individual from practices that apparently appear to be illogical. Moreover, the results of beliefs show that an increase in education makes the beliefs milder. Therefore, further exploration is required to understand this pattern related to education.

Coefficients of occupation are not significant, which implies that occupation does not influence taboos. However, if we consider significance level at 10 per cent,[4] the college-going girls are found to observe stricter taboos.

The coefficient of taboos for women in joint family is not statistically significant, whereas those in nuclear family have a positive and statistically significant coefficient. Thus, women in joint and extended families observe milder taboos compared to those in nuclear families.

The coefficients of taboos score for those living in developing as well as tribal districts are higher and statistically significant. Thus, those living in developing and tribal districts have milder taboos compared to the developed ones.

Table 2.5 Taboo Score Statistics

Statistics	Taboos Score
Min	−0.5807
1st quartile	−0.1224
Median	0.0026
Mean	0
3rd quartile	0.1693
Max.	0.4193

The results of the regression reveal that except for Sikhs, women from all other religions practice stricter taboos as compared to Christian women. Urban women practice lesser/milder taboos. The practice of taboos gets milder with education. College-going girls are found to be practicing stricter taboos than women across occupations, including school-going girls. Women practice from joint and extended families is practice lesser/milder taboos compared to those from nuclear families. The practices of taboos by the women living in developing or tribal districts practice milder taboos.

2.3 Key Findings

- Most girls received information and guidance about menarche and initial menstrual cycles from their mothers.
- Most women were aware about the physical changes and hormonal changes around menarche. However, there was low awareness among women about a medical condition that would manifest around menarche. Awareness about the source of menstrual blood was also relatively low.
- Awareness is lower among girls/women living in rural areas, developing, and tribal districts. Beliefs are milder, and less taboos are practiced in rural areas and developed districts.
- Most women do not hold beliefs surrounding menstruation but continue to practice taboos because their minds are conditioned since childhood.
- Awareness is lower among non-working women. Also, non-working women seem to hold stronger beliefs around menstruation and menstrual cycles.
- Awareness increases with the level of education. Also, the beliefs get milder with education but not the taboos. There is no influence of occupation on practicing of taboos.
- Marital status does not influence awareness.

Table 2.6 Impact of Socioeconomic Factors on Taboos Practiced during Menstruation

Socioeconomic Factors*	Estimate	Std. Error	t-value	Pr(>|t|)
(Intercept)	0.0675	0.0632	1.069	0.2855
1. Religion: Hindu	−0.1223	0.0410	−2.98	0.0030
Jain	−0.4274	0.0510	−8.379	<0.0001
Muslim	−0.1332	0.0591	−2.256	0.0243
Parsi	−0.2479	0.1503	−1.649	0.0994
Sikh	−0.1480	0.1510	−0.98	0.3272
2. Region: Urban	0.0785	0.0147	5.329	<0.0001
3. Education: Illiterate	−0.2854	0.0551	−5.172	<0.0001
Literate (with or without schooling)	0.0151	0.0386	0.39	0.6963
Up to 12th grade	−0.0336	0.0207	−1.62	0.1056
Up to primary	−0.0957	0.0314	−3.048	0.0024
Up to secondary	−0.0915	0.0240	−3.824	0.0001
4. Occupation: Working (indoors)	−0.0120	0.0333	−0.36	0.7189
Other work (outdoors)	0.0548	0.0341	1.605	0.1089
College-going student	−0.0579	0.0341	−1.701	0.0893
Non-working	−0.0361	0.0344	−1.05	0.2941
School-going student	−0.0462	0.0418	−1.106	0.2692
5. Family type: Joint	0.0331	0.0364	0.911	0.3625
Nuclear	0.0757	0.0367	2.067	0.0390
6. Development status: Developing	0.0331	0.0171	1.936	0.0531
Tribal	0.0530	0.0217	2.438	0.0149
Undefined	0.0249	0.0567	0.439	0.6605

Residual standard error: 0.2041 on 1003 degrees of freedom
Multiple R-squared: 0.2192, Adjusted R-squared: 0.2029
F-statistic: 13.41 on 14 and 1003 df, p-value: < 0.0001

* The reference category for Religion is "Christians". The reference categories for rest of the variables are same as for the regressions of Table 2.2.

- Religion does not seem to be an important component to explain awareness or beliefs in our models. However, women belonging to Sikhs seem to be practicing the least taboos as compared to all other religions.

Since awareness, beliefs, and taboos influence the choice of menstrual products and hygiene habits, we discuss the cost scenarios for sanitary napkins/tampons and menstrual cups in the next chapter. We also discuss the hygiene scenarios and the costs associated with them in chapters 3 and 4.

Notes

1 "Saturation" is a concept in qualitative research and it is defined by Glaser and Strauss (2017) as "… no additional data are being found whereby the sociologist can develop properties of the category." (Saunders et al., 2018).
2 It may be noted that unlike other studies in healthcare, menstrual health does not involve patients, because menstruation is neither a disease or a specific condition, nor does it require treatment under normal circumstances. Therefore, the word patient is not applicable. But we codesigned this questionnaire with girls/women in reproductive span.
3 A forward stepwise regression model begins with no predictors/explanatory variables and adds the one that is the most statistically significant at each step, until there are no statistically significant variables. One may find details of many such algorithms, including forward stepwise, with their pros and cons in Khaire and Dhanalakshmi (2022). Forward selection is the most commonly used algorithms of all (Smith, 1990). For more details on forward stepwise regression, see technical notes in Appendix F.
4 In social science research, the normally accepted significance levels are 1 and 5 per cent. However, researchers may, depending upon the nature of study, set the significance level to 10 per cent (Doan, 2005).

References

Abdul Hakim, N. S., Omar, M. A., & Abdul Hakim, M. H. (2024). Knowledge and practices on menstruation among adolescent girls in rural areas in Malaysia. *International Journal of Advanced Research in Future Ready Learning and Education*, *32*(1), 1–7. https://doi.org/10.37934/frle.32.1.17

Afiaz, A., & Biswas, R. K. (2021). Awareness on menstrual hygiene management in Bangladesh and the possibilities of media interventions: Using a nationwide cross-sectional survey. *BMJ Open*, *11*(4), e042134. https://doi.org/10.1136/bmjopen-2020-042134

Al Mutairi, H., & Jahan, S. (2021). Knowledge and practice of self-hygiene during menstruation among female adolescent students in Buraidah city. *Journal of Family Medicine and Primary Care*, *10*(4). https://journals.lww.com/jfmpc/fulltext/2021/10040/knowledge_and_practice_of_self_hygiene_during.13.aspx

Anusree, P. C., Roy, A., Sara, A. P., Faseela, V. C. M., Babu, G. P., & Tamrakar, A. (2014). Knowledge regarding menstrual hygiene among adolescent girls in selected school, Mangalore with a view to develop an information booklet. *IOSR Journal of Nursing and Health Science*, *3*(1), 55–60.

Apffel-Marglin, F. (2020). Let's break the vicious circle of menstrual taboo. *The Sacred Groves*. https://www.manushi.in/gender-justice/-the-sacred-groves

Arshad Ali, S., Baloch, M., Riaz, L., Iqbal, A., Riaz, R., Perveen, B., Siddiqui, M., & Arshad Ali, A. (2020). Perceptions, practices, and challenges regarding menstrual hygiene among women in Karachi, Pakistan: A comparison between general population and healthcare workers. *Cureus, 12*(8), e9894. https://doi.org/10.7759/cureus.9894

Arumugam, B., Nagalingam, S., Mahendra Varman, P., Ravi, P., & Ganeshan, R. (2014). Menstrual hygiene practices: Is it practically impractical? *International Journal of Medicine and Public Health, 4*(4), 472–476.

Ashley-Montagu, M. F. (1940). Physiology and the origins of the menstrual prohibitions. *The Quarterly Review of Biology, 15*(2), 211–220. JSTOR.

Brook, C. G. D., Jacobs, H. S., & Stanhope, R. (1988). Polycystic ovaries in childhood. *British Medical Journal (Clinical Research Edition), 296*(6626), 878–878. JSTOR.

Bulto, G. A. (2021). Knowledge on menstruation and practice of menstrual hygiene management among school adolescent girls in Central Ethiopia: A cross-sectional study. *Risk Management and Healthcare Policy, 14*, 911–923. https://doi.org/10.2147/RMHP.S296670

Chadalawala, U. R., & Kala, S. (2016). Assessment of menstrual hygiene practices among adolescent girls. *Stanley Medical Journal, 3*(1), 13–16.

Crawford, P. (1981). Attitudes to menstruation in seventeenth-century England. *Past & Present, 91*, 47–73. JSTOR.

Darrow, S. L., Vena, J. E., Batt, R. E., Zielezny, M. A., Michalek, A. M., & Selman, S. (1993). Menstrual cycle characteristics and the risk of endometriosis. *Epidemiology, 4*(2), 135–142. JSTOR.

Das, M. (2008). Menstruation as pollution: Taboos in Simlitola, Assam. *Indian Anthropologist, 38*(2), 29–42.

Deo, D. S., & Ghattargi, C. H. (2005). Perceptions and practices regarding menstruation: A comparative study in urban and rural adolescent girls. *Indian Journal of Community Medicine, 30*(1), 33–34.

Doan, A. E. (2005). Type I and Type II error. In K. Kempf-Leonard (Ed.), *Encyclopedia of social measurement* (pp. 883–888). Elsevier. https://doi.org/10.1016/B0-12-369398-5/00110-9

Dündar, T., & Aksu, H. (2022). Cultural beliefs and practices of reproductive women about menstruation. *Kadın Sağlığı Hemşireliği Dergisi, 8*(2), 41–49.

Engemise, S., Gordon, C., & Konje, J. C. (2010). EASILY MISSED: Endometriosis. *BMJ: British Medical Journal, 340*(7761), 1414–1415. JSTOR.

Ganguly, A., Ganguly, M., Jana, S., & Midya, D. K. (2024). Menstruation in adolescent girls: Myths & taboos. *Acta Biologica Slovenica, 67*(2), 64–74.

Ganguly, L., Satpati, L., & Nath, S. (2021). "Taboos and Myth" – Indispensable part of menstruation: An overview. *Asian Pacific Journal of Health Sciences, 8*(4), 250–253. https://doi.org/10.21276/apjhs.2021.8.4.28

Garg, S., Sharma, N., & Sahay, R. (2001). Socio-cultural aspects of menstruation in an urban slum in Delhi India. *Reproductive Health Matters, 9*(17), 16–25.

Glaser, B., & Strauss, A. (2017). *Discovery of grounded theory: Strategies for qualitative research*. Routledge.

Gori, M., Betti, A., & Melacci, S. (2024). Chapter 2—Learning principles. In M. Gori, A. Betti, & S. Melacci (Eds.), *Machine learning* (2nd ed., pp. 53–111). Morgan Kaufmann. https://doi.org/10.1016/B978-0-32-389859-1.00009-X

Gujarati, D. N., Porter, D. C., & Gunasekar, S. (2011). *Basic econometrics* (5th ed.). Mcgraw Hill Education (India) Private Limited.

Kaur, R., Kaur, K., & Kaur, R. (2018). Menstrual hygiene, management, and waste disposal: Practices and challenges faced by girls/women of developing countries. *Hindawi Journal of Environmental and Public Health*, Online. https://doi.org/10.1155/2018/1730964

Khaire, U. M., & Dhanalakshmi, R. (2022). Stability of feature selection algorithm: A review. *Journal of King Saud University - Computer and Information Sciences*, *34*(4), 1060–1073. https://doi.org/10.1016/j.jksuci.2019.06.012

Khan, H. N., Digal, G., & Wani, M. A. (2024). Women, health and marginality: A menstruation practices, beliefs and taboos in border villages of Jammu And Kashmir. *Educational Administration: Theory and Practice*, *30*(5), 11223–11230. https://doi.org/10.53555/kuey.v30i5.4929

Kotian, S. M., Chaudhary, V. K., Mutya, V. S., Sekhon, A. S., Sriraman, S., & Prasad, P. (2017). Assessment of knowledge, practice and perception of menstruation among adult women in the reproductive age group, in Mangalore, India. *International Journal of Reproduction, Contraception, Obstetrics and Gynecology*, *6*(10), 4595–4601.

Krishnan, S., & Twigg, J. (2016). Menstrual hygiene: A 'silent' need during disaster recovery. *Waterlines*, *35*(3), 265–276. https://doi.org/10.3362/1756-3488.2016.020

Kumar, A., Dhadwal, Y., Yadav, V., & Sharma, B. (2024). A cross-sectional study of knowledge, taboos, and attitudes towards menstruation. *Ethnicity & Health*, *29*(2), 208–219. https://doi.org/10.1080/13557858.2023.2293450

Kumar, A., & Srivastava, K. (2011). Cultural and social practices regarding menstruation among adolescent girls. *Social Work in Public Health*, *26*(6), 594–604. http://dx.doi.org/10.1080/19371918.2010.525144

Kumari, S., Sood, S., Davis, S., & Chaudhury, S. (2021). Knowledge and practices related to menstruation among tribal adolescent girls. *Industrial Psychiatry Journal*, *30*(Suppl 1). https://journals.lww.com/inpj/fulltext/2021/30001/knowledge_and_practices_related_to_menstruation.30.aspx

Lacroix, A., Gondal, H., & Langaker, M. (2021). *Physiology, Menarche*. StatPearls. https://www.ncbi.nlm.nih.gov/books/NBK470216/

Mahajan, A., & Kaushal, K. (2017). A Descriptive study to assess the knowledge and practice regarding menstrual hygiene among adolescent girls of government school of Shimla, Himachal Pradesh. *Journal of Health and Research*, *4*(2), 99–103.

Mahon, T., & Fernandes, M. (2010). Menstrual hygiene in South Asia: A neglected issue for WASH (water, sanitation and hygiene) programmes. *Gender and Development*, *18*(1), 99–113. JSTOR.

Michael, J., Iqbal, Q., Haider, S., Khalid, A., Haque, N., Ishaq, R., Saleem, F., Hassali, M. A., & Bashaar, M. (2020). Knowledge and practice of adolescent females about menstruation and menstruation hygiene visiting a public healthcare institute of Quetta, Pakistan. *BMC Women's Health, 20*(1), 4. https://doi.org/10.1186/s12905-019-0874-3

Montgomery, R. E. (1974). A cross-cultural study of menstruation, menstrual taboos, and related social variables. *Ethos, 2*(2), 137–170. JSTOR.

Mukherjee, A., Lama, M., Khakurel, U., Jha, A. N., Ajose, F., Acharya, S., Tymes-Wilbekin, K., Sommer, M., Jolly, P. E., Lhaki, P., & Shrestha, S. (2020). Perception and practices of menstruation restrictions among urban adolescent girls and women in Nepal: A cross-sectional survey. *Reproductive Health, 17*(1), 81. https://doi.org/10.1186/s12978-020-00935-6

Neupane, M. S., Sharma, K., Bista, A. P., Subedi, S., & Lamichhane, S. (2020). Knowledge on menstruation and menstrual hygiene practices among adolescent girls of selected schools, Chitwan. *Journal of Chitwan Medical College, 10*(1), 69–73.

NIPCCD. (2014). *Improvement in knowledge and practices of adolescent girls regarding reproductive health with special emphasis on hygiene during menstruation in five years.* National Institute of Public Cooperation and Child Development.

Patavegar, B. N., Kapilashrami, M. C., Rasheed, N., & Pathak, R. (2014). Menstrual hygiene among adolescent school girls: An in-depth cross-sectional study in an urban community. *International Journal of Health Sciences & Research, 4*(11), 15–21.

Prajapati, J., & Patel, R. (2015). Menstrual hygiene among adolescent girls: A cross-sectional study in urban community of Gandhinagar. *The Journal of Medical Research, 1*(4), 122–125.

Robeva, R., & Kumanov, P. (2016). Physical changes during pubertal transition. In P. Kumanov & A. Agarwal (Eds.), *Puberty: Physiology and abnormalities* (pp. 39–64). Springer International Publishing. https://doi.org/10.1007/978-3-319-32122-6_4

Sarkar, I., Dobe, M., Dasgupta, A., Basu, R., & Shahbanu, B. (2017). Determinants of menstrual hygiene among school going adolescent girls in a rural area of West Bengal. *Journal of Family Medicine and Primary Care, 6*(3), 583–588.

Saunders, B., Sim, J., Kingstone, T., Baker, S., Waterfield, J., Bartlam, B., Burroughs, H., & Jinks, C. (2018). Saturation in qualitative research: Exploring its conceptualization and operationalization. *Quality & Quantity, 52*(4), 1893–1907. https://doi.org/10.1007/s11135-017-0574-8

Smith, P. (1990). The use of performance indicators in the public sector. *Journal of the Royal Statistical Society. Series A (Statistics in Society), 153*(1), 53–72. https://doi.org/10.2307/2983096

Sreenivasa, N. S., Sakranaik, S., Sobagiah, R. T., & Viswanath. (2017). Perception of menstruation and practices among adolescents in urban field practice area, Bangalore: A cross sectional study. *National Journal of Community Medicine, 8*(11), 645–649.

Sridhar, N. (2019). *Menstruation across cultures: A historical perspective.* Vitasta Publishing Pvt Ltd.

Srinivasa, S., & Manasa, G. (2017). Menstrual hygiene among school-going adolescent girls. *TJPRC: International Journal of General Pediatrics and Medicine, 2*(2), 17–22.

Stephens, W. N. (1961). A cross-cultural study of menstrual taboos. *Genetic Psychology Monogprahs, 64,* 385–416.

Strange, J.-M. (2000). Menstrual fictions: Languages of medicine and menstruation, c. 1850–1930. *Women's History Review, 9*(3), 607–628. https://doi.org/10.1080/09612020000200260

Syed Abdullah, S. Z. (2022). Menstrual food restrictions and taboos: A qualitative study on rural, resettlement and urban indigenous Temiar of Malaysia. *PLoS ONE, 17*(12), e0279629. https://doi.org/10.1371/journal.pone.0279629

Thakur, H., Arosson, A., Bansode, S., Lundborg, C. S., Dalvie, S., & Faxelid, E. (2014). Knowledge, practices, and restrictions related to menstruation among young women from low socioeconomic community in Mumbai, India. *Frontiers in Public Health, 2,* 2–7.

Uppal, M. K., Rana, M., & Batta, A. (2022). Taboos related to menstruation in various cultures. *Journal of Pharmaceutical Negative Results, 13*(9), 8451–8456.

van Eijk, A. M., Sivakami, M., Bora Thakkar, M., Bauman, A., Laserson, K. F., Coates, S., & Phillips-Howard, P. A. (2016). Menstrual hygiene management among adolescent girls in India: A systematic review and meta-analysis. *BMJ Open, 6.* http://dx.doi.org/10.1136/bmjopen-2015-010290

3 Direct Costs of Menstrual Hygiene Practices

Choice of menstrual products is crucial to menstrual hygiene. Economics explains how choices are made. However, standard classical economics theory suggests that a rational woman would optimise her choice (and the resultant utility), subject to budget constraint. Choices are however, constrained by the information asymmetry and access to the products, given the market imperfections. Choice of menstrual product could be better explained by behavioural economics. Theories of behavioural economics suggest that individuals do not always attempt at maximising their utility; rather, they may maximise the utility of others within their budget constraints. Such choices are termed as "other-regarding preferences" (Fehr & Schmidt, 2006). Social conditioning of a woman drives her choice in favour of the products for family consumption over those for her individual consumption, given her budget constraint, and the choice of menstrual products is no exception.

Menstruation is one such health condition that is experienced by a woman every 28 days, for almost a two-third of her life. The inner lining of the uterus is shed, which results in bleeding. This blood requires to be soaked by some absorbent. Typically, a woman sheds 60 millilitres (2.7 ounces) of blood, that is equivalent to one and a half shot glasses full, during one menstrual cycle (InformedHealth.org [Internet], 2017). Moreover, it is important to look after the vaginal hygiene during menstruation since, it becomes more susceptible to infections during the menstrual cycles. Therefore, the choice of product and the way it is used to soak blood are important to ensure vaginal hygiene.

Information asymmetry about menstrual products arises out of the limited choice set comprising the products available in the vicinity of the girl's/woman's residence. Education, media, and the activities of non-governmental organisations (NGOs) help in reducing this information asymmetry to a great extent. Many small and large-scale NGOs have been active in spreading awareness about menstruation, menstrual products, and hygiene associated with menstrual product practices. They also work towards selling menstrual hygiene products, mainly sanitary napkins,

DOI: 10.4324/9781003473961-3

either at subsidised rates or for free. In some areas, especially in the rural areas, even government institutions like Anganwadis[1] and ASHA[2] workers help reduce information asymmetry and increase the affordability (by providing the sanitary napkins at subsidised rates). Online shopping platforms also help in reducing the information asymmetry and also in increasing access to a wide range of menstrual products. However, online platforms cater only to those women who have access to smartphones and internet. Also, online buying requires planning in advance. So, in case of an emergency requirement of an absorbent, a nearby local store appears to be the most convenient place to buy the menstrual product. Online orders are restricted by the locations served by those online selling platforms (like Amazon, Flipkart, BigBasket, etc.).

In cases where the menstrual products are sold at subsidised rates by the government or the NGO, the state/NGO bears that proportion of the cost. In the absence of availability of any such subsidised menstrual products, the entire cost is required to be borne by an individual (the woman/ the girl). Therefore, an effort is made to understand and estimate the costs of menstrual hygiene, as it would give insights into an individual's affordability of menstrual products. This estimation can also give insights to the policymakers about pricing the product, determining the subsidy amount, or making it available for free to women of lower socioeconomic strata. Estimating the explicit cost of a healthcare product can be helpful to the policymakers in terms of subsidising the products, making available different products at subsidised rates, developing an awareness campaign, etc.

The choice of menstrual product constitutes the direct cost. Menstrual hygiene practices like regular cleaning of the private parts and thorough washing of hands after using the washroom also add to direct costs. MHM practices require using at least soap, though women with broader choices may use intimate hygiene products to clean the private parts and disinfectant/hand wash/hand sanitiser to wash the hands. The cost of these products adds to the direct costs of menstrual hygiene. On the other hand, non-adherence to hygiene can result in infections and diseases, and menstruation is no exception.

This chapter discusses the practices adopted to ensure menstrual hygiene, including the products used to soak menstrual blood and the products used for cleaning/disinfecting the private parts and hands. The methodology of estimation of direct costs, to ensure menstrual hygiene is discussed in this chapter. This chapter is organised in five sections: Section 3.1 describes the patterns in menstrual hygiene practices, especially the awareness and usage patterns of menstrual products. This section also examines whether the socioeconomic background has any influence on the choice of menstrual products. While Chapter 2 discusses awareness, surrounding menstruation, the focus of this section is on product awareness and usage patterns. Section 3.2 describes the product preferences and the place from where they purchase, whether they buy on their own or somebody else buys it for them, whether they hesitate in asking for menstrual products, who accompanies them while they go to purchase the products, and the like. This allied information is important as it is likely to influence the product preference, which, in turn, influences the costs. Section 3.3 presents the methodology to estimate the annual direct cost of menstrual products and also provides an estimate of the same based on our survey data. Section 3.4 gives the methodology of estimation of costs of allied products to ensure menstrual hygiene and provides an annual estimate for the same. Section 3.5 discusses the prevalence of medical conditions and their implications on the regularity and consistency of menstrual cycles. The methodology of estimating the annual medical costs to treat these issues, which constitute cost components, is described in this section. Women with medical conditions or menstrual dysfunction, without any medical conditions, also undergo psychological shock/depression. The implication on medical costs resulting from psychological disturbances is also discussed in this section, though they constitute indirect costs. Section 3.6 summarises the findings on direct costs. It may be noted that we have not attempted to estimate the imputed costs.

3.1 Menstrual Hygiene Practices

The extent of information asymmetry is reflected by examining the awareness of women about the varieties of menstrual products available in the market. As discussed earlier, information asymmetry and availability of products in the neighbourhood are the determining factors for the choice of menstrual products, over and above budget constraints. This section examines the extent of awareness of respondents about the menstrual products available in the market, and the pattern in menstrual products used by them.

Figure 3.1(a) shows the awareness of the respondents, regarding the products available for soaking menstrual blood. It is likely that any respondent would be aware about more than one menstrual product, and hence the total percentages exceed 100 per cent. Figure 3.1(b) shows the percentage of respondents who are aware of the disposable sanitary napkins being made available by the government at subsidised rates.

Figure 3.1(a) shows that there is almost equal awareness about disposable sanitary napkins (45.75 per cent), cloth (43.70 per cent) (from old sarees), and cotton pads (47.12 per cent). There is relatively lower awareness about menstrual cups (19.51 per cent) and tampons (16 per cent). The pads made from medicated cotton are known as cotton pads. It may be noted that even some disposable sanitary napkins, which are made from compressed cotton, are also cotton pads. Thus, all

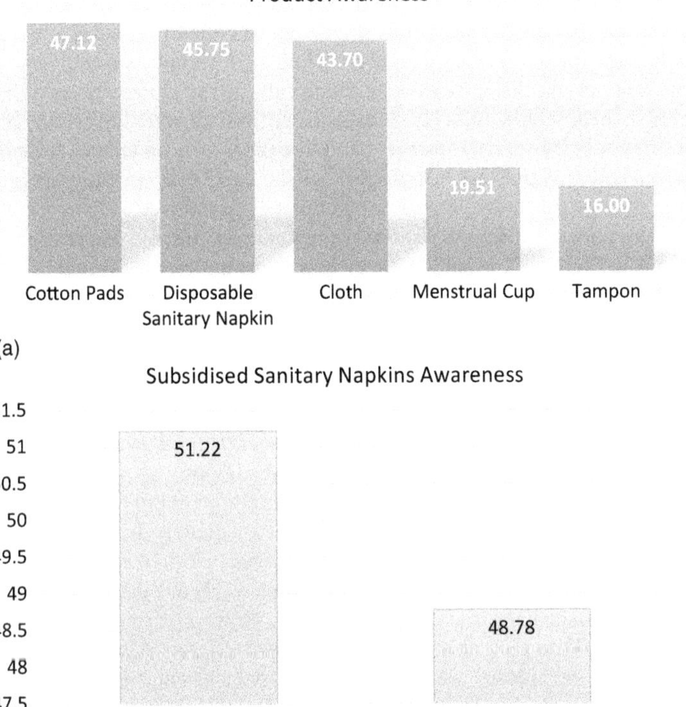

Figure 3.1 Awareness about Menstrual Products.

these categories of cotton pads are included for the option "cotton pads" in examining the product awareness. These cotton pads could be disposable or reusable.

Figure 3.1(b) shows that more than 50 per cent respondents, out of 1025, are aware that the government is distributing disposable sanitary napkins at subsidised rates. Higher awareness is likely to result in more women making use of the disposable sanitary napkins, instead of traditional absorbents like cow dung, mud, leaves, or even cloths from old sarees. One cannot say with certainty that cloth from old sarees used as absorbents is less hygienic than the sanitary napkins/tampons because the latter contain chemicals to transform blood into gel, which helps in enhancing the blood retention capacity. This in turn, would add to the convenience in terms of requiring replacing the sanitary napkins with lesser frequency, in comparison with that of cloths from old sarees. The absorbent in the case of disposable sanitary napkins is made up of super-absorbent polymer (made from sodium polyacrylate) granules and fluff cellulose, which are encapsulated by cellulose or (nonwoven) polypropyl-ene (Bae et al., 2018; Woeller & Hochwalt, 2015). Polymeric materials used in disposable sanitary napkins are identified as potential carcinogens (Earls et al., 2003). On the other hand, the problem with clothes from old sarees is that if they are not properly washed and dried in open space or sun, they would be smelly and not completely disinfected.

As against, if disposable sanitary napkins and tampons are changed at stipulated intervals, usually between four to six hours, they would be rela-tively hygienic. However, if these are used longer than six hours, they could result in rashes and vaginal infections, as dampness due to excess blood results in the breeding of bacteria and fungus. Keeping tampons for longer hours could cause rashes while pulling them out since they would be drenched with menstrual blood; in extreme cases, it could result in toxic shock syndrome (TSS)[3] (Santos-Longhurst, 2020). However, assuming the use of disposable sanitary napkins and tampons as per the hygiene norms, they turn out to be relatively hygienic compared to cloths from old sarees that are reused after washing and not adequately dried in the sun.

The pattern in place of drying the absorbent cloth, torn from an old saree, is given in Figure 3.2, which shows that nearly 60 per cent of women dry their washed absorbent cloth in the sun. However, around 37 per cent of them still dry their absorbent cloth inside the house. Also, if the deter-gent is not completely washed out, it causes rashes and irritation while reusing these cloths. This makes the use of cloths from old sarees rela-tively more vulnerable to vaginal itching and infections.

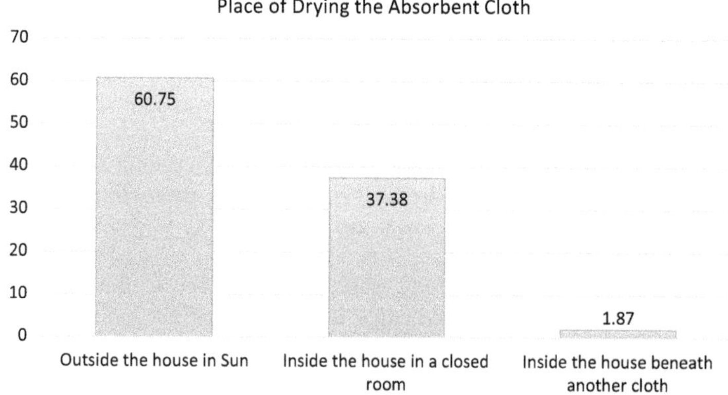

Figure 3.2 Place of Drying the Absorbent Cloth Torn from an Old Saree.

Figure 3.3 gives details of the product that is normally/regularly used by the respondents during their menstrual cycles to soak blood. During the data collection, the respondents were asked a counter-question that they use more than two products, one as the main source to soak blood and the other to prevent leakage. In such situations, the respondents were asked to mention the product used as the main source as their normally used product.

Figure 3.3 shows that, a majority (58.34 per cent) of respondents use disposable sanitary napkins. Though, 10.44 per cent of women use cloth from old sarees/old cloth as absorbent. The users of menstrual cups and tampons users are very few (around one per cent). However, in our sample of 1025, we found one woman (0.10 per cent) women who used natural materials like mud/cow dung/leaves to soak menstrual blood and 2 women (0.20 per cent) who used nothing to soak menstrual blood. These are those women/girls who are trained by their mothers to hold menstrual blood from flowing, and they regularly go to the toilets/places for urination during their period, to discharge their menstrual blood.

It is hypothesised that usage pattern for the menstrual products depend on exposure to products, together with the amount of personal disposable income. The exposure to products could be associated with the education, nature of occupation, region (rural/urban), and the extent of development of the districts (developed, developing, or tribal districts) to which the respondents belong.

We have classified menstrual cups, disposable sanitary napkins, and tampons as "high hygiene value" products and the rest as "low hygiene value". We have used only economic indicators as predictors.

A binary logistic regression examines the influence of exposure and income on the choice of high versus low hygiene value products. The results of these two regressions are given in Table 3.1:

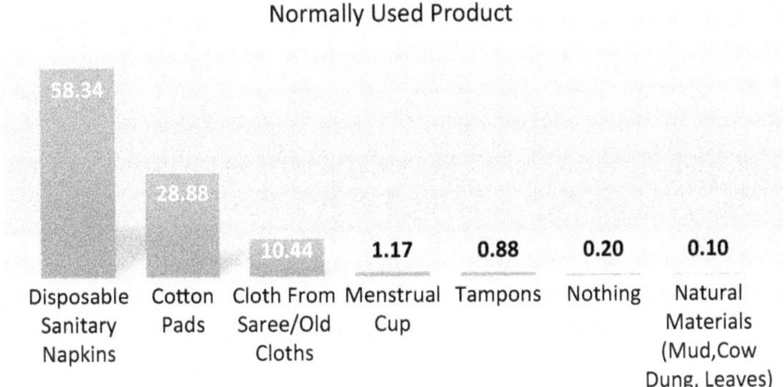

Figure 3.3 Product Normally Used to Soak Menstrual Blood.

The "Estimates" in Table 3.1 are the ln(odds) of "high hygiene value". These are converted to odds by taking an exponent of the ln(odds) and they are further converted to using Equation 3.1:

$$\frac{odds}{(1 + odds)}$$

(3.1)

The probability estimates given in Table 3.1 show that:

Compared to the women in high-income category, the probability of using high hygiene value products by those belonging to the upper and lower middle-income groups is higher. There is no statistically significant difference between high- and low-income categories, in terms of the probability of using high hygiene value products. Information asymmetry is reducing fast because of smartphones and access to mass media. *Albeit*, those from low-income group would have access to the disposable sanitary napkins made available by the government at subsidised rates. This population is also served by NGOs working for menstrual hygiene, and therefore, they might have better access to high hygiene value products, especially to disposable sanitary napkins, in comparison with those who belong to the middle-income group.

The probability estimates and the p-values associated with the Ln(odds) of education suggest that in comparison with those who have attained higher education, the probability of using high hygiene value products for all other levels of education is low (between 0.16 and 0.37). The coefficient of the illiterate category of respondents is statistically insignificant. This indicates that there is no difference in probability of use of high hygiene value products by those who have attained higher education and those who are illiterate. The respondents who are illiterate are more likely to belong to low-income group. Therefore, they might have better access to disposable sanitary napkins, either to the ones sold at subsidised rates by the government or to those made available by NGOs. Some NGOs sell disposable sanitary napkins at subsidised rates, whereas some others distribute them free of cost. In either case, this creates awareness about disposable sanitary napkins among the deprived and vulnerable.

The Ln(odds) of women in all categories except college-going girls are insignificant. This implies that only college-going girls have a 0.2664 lower probability of using the high hygiene value products as compared to women in all other occupation categories. The reason for this anomalous result remains to be explained.

The probability of women from tribal districts using high hygiene products as absorbents is low by 0.3565 as compared to those from developed and developing districts. However, there is no difference across rural

Table 3.1 Determinants of Menstrual Hygiene Products Preference

Economic Factors*	Estimate Ln(Odds)	Std. Error	z-Value	Exp (Estimate) Odds	Probability Odds/ (1 + Odds)	Pr (>|z|)
(Intercept)	0.7887	0.4608	1.712	2.20051	0.6876	0.0869
1. Income category: low	0.8455	0.4701	1.799	2.32921	0.6996	0.0721
Lower-middle	0.5469	0.2889	1.893	1.72794	0.6334	0.0583
Upper-middle	0.5985	0.2796	2.141	1.81939	0.6453	0.0323
2. Education: illiterate	−0.4921	0.5780	−0.851	0.61137	0.3794	0.3946
Literate (with or without schooling)	−1.6503	0.4434	−3.722	0.19199	0.1611	0.0002
Up to 12th grade	−0.5285	0.2165	−2.440	0.58952	0.3709	0.0147
Up to primary	−1.4771	0.3348	−4.412	0.22829	0.1859	<0.0001
Up to secondary	−1.4218	0.2536	−5.606	0.24128	0.1944	<0.0001
3. Occupation: working indoors	−0.5178	0.3812	−1.358	0.59581	0.3734	0.1743
Other work – outdoors	−0.3100	0.3917	−0.791	0.73345	0.4231	0.4287
College-going student	−1.0129	0.3899	−2.598	0.36315	0.2664	0.0094
Non-working	−0.4731	0.3914	−1.209	0.62309	0.3839	0.2267
School going student	−0.0755	0.4591	−0.164	0.92727	0.4811	0.8694
4. Region: urban	0.1427	0.1566	0.911	1.15334	0.5356	0.3623
5. Development status: developing	−0.1064	0.1850	−0.575	0.89909	0.4734	0.5652
Tribal	−0.5905	0.2392	−2.469	0.55408	0.3565	0.0136
Undefined	0.2158	0.6446	0.335	1.24082	0.5537	0.7378

Null deviance: 1317.4 on 975 degrees of freedom
Residual deviance: 1220.5 on 958 degrees of freedom
49 observations deleted due to missingness
AIC: 1256.5 Number of fisher scoring iterations: 4

* For categorical predictors like income, education, occupation, region, and development status, the coefficients are interpreted in the context of a reference category. In case of categorical predictors, the coefficients will be calculated for all the categories, except the reference category. The reference category for income is "High" income group and for occupation is "Working outdoor for manual labour". The reference category for region is "rural" and that for development status is "developed".

and urban areas in terms of usage of high hygiene products. This could also be because of active involvement of NGOs, Anaganwadi, and ASHA in rural areas.

On one hand, there are 0.20 per cent, who do not use anything to soak menstrual blood (Figure 3.3), there are 32 per cent of women who use more than one product (Figure 3.4 a) during their menstrual cycles.

A detailed examination of those 330 (32 per cent) respondents (Figure 3.4 b), who use more than one product, shows that close to 80.30 per cent use cotton pads and use disposable sanitary napkins. Out of these 330 women, 4.84 use natural materials. Since these women use more than one product, the percentage of users of different products exceeds 100 per cent.

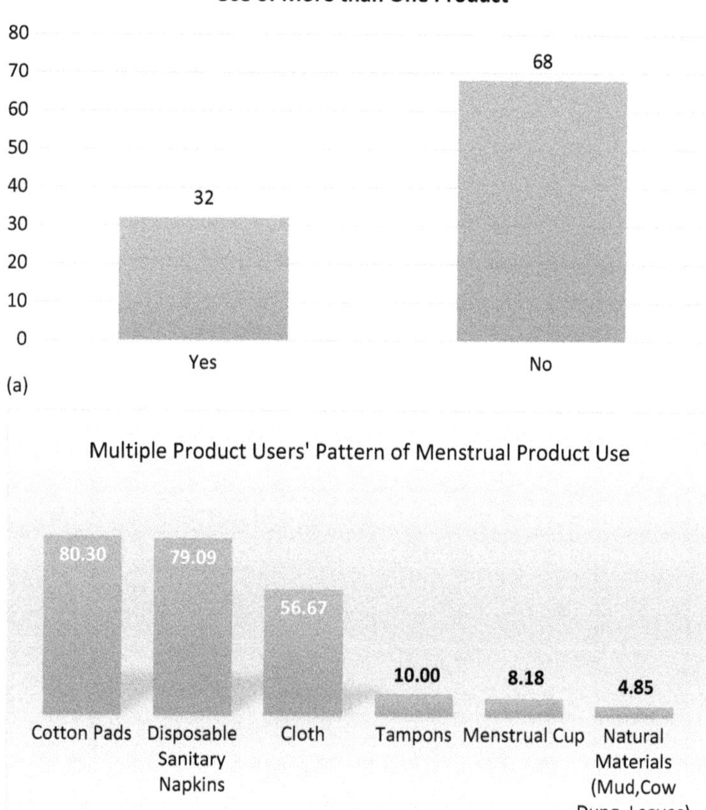

Figure 3.4 Women using More than One Menstrual Products.

3.2 Menstrual Product Preferences

The preference for a particular brand or product can be examined by understanding the reasons for change in brands and products. Out of 1025 respondents, 410 respondents have changed brands for their primary products, the normally used products. Since there are 330 women who use more than one product during their menstrual cycles, the term "primary product" is used for the main absorbent out of two or more products used by them; the other one is termed as "secondary product". Out of 330 respondents who use more than one product, 147 have changed the brands of their secondary products. Also, out of 1025 respondents, 279 have even switched over from one menstrual product used for soaking blood to another. The reasons for change in brands and change in products could be found in Figure 3.5. The pattern of transition from one product to another can also be seen in Figure 3.6.

Figure 3.5 (a) shows that women change brands mainly for better quality; 37.85 per cent of respondents have changed the brands of their primary product and 28.48 per cent have changed the brands of their secondary product, and that is because the newer brand offers better quality products. The second most cited reason for brand change of the primary product by the respondents is increased comfort (26.24 per cent) and for the secondary product, it is because the new brand offers more environmentally friendly products (20.30 per cent). The most cited reason why women change primary products (Figure 3.5 b) is for improved hygiene (67.74 per cent) and for the secondary product, it is increased comfort (53.41 per cent). These percentages are calculated from 1025 women for primary (or the only) product, and from 330 for the secondary product.

Figure 3.6 shows that a large chunk of the respondents (46.74 per cent) have switched over from the cloth torn from an old saree to using disposable sanitary napkins. This is followed by a switch over from cotton pads to disposable sanitary napkins (11.59 per cent) and from locally prepared napkins (could be cotton-based or cloth-based) to disposable sanitary napkins and cotton pads (9.42 per cent and 7.25 per cent respectively). Respondents have also switched over from using nothing or natural materials like mud, dung, etc., to other hygienic options.

Even today, menstruation is not discussed openly and is certainly uncommon to discuss with men. Information asymmetry also influences the choice of product. Therefore, most women have limited exposure to the menstrual products that are available in the market. Data on the outlet from which they buy the menstrual product determines the extent of information available to them. Also, whether the girl/woman buys the product on her own or someone on her behalf buys the product also determines her information set about the products and, thereby, her choice of product. Therefore, the information on place of purchase, whether she

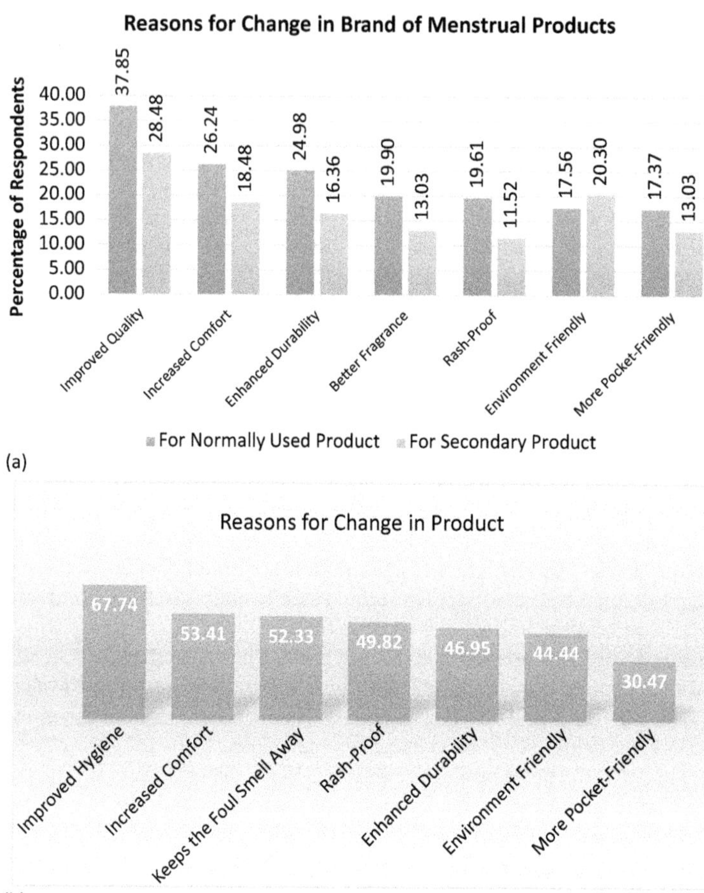

(a)

(b)

Figure 3.5 Reasons for Product Preference as Captured by Change in Brand/ Product.

purchases by herself or not, whether her family members/husband (only married women) have any say in the product to be used, and whether she hesitates to buy the product, etc., is sought and presented in Figure 3.7.

A majority of the respondents (57.52 per cent), as shown in Figure 3.7(a), buy the menstrual products from a shop in the nearby area. This is partly because in small towns and rural areas, there are no options of departmental stores/superstores. Therefore, the choice is confined to local area shops or medical stores. Women in bigger cities seem to prefer buying from the departmental store/superstore. In our survey, 37.98 per cent of

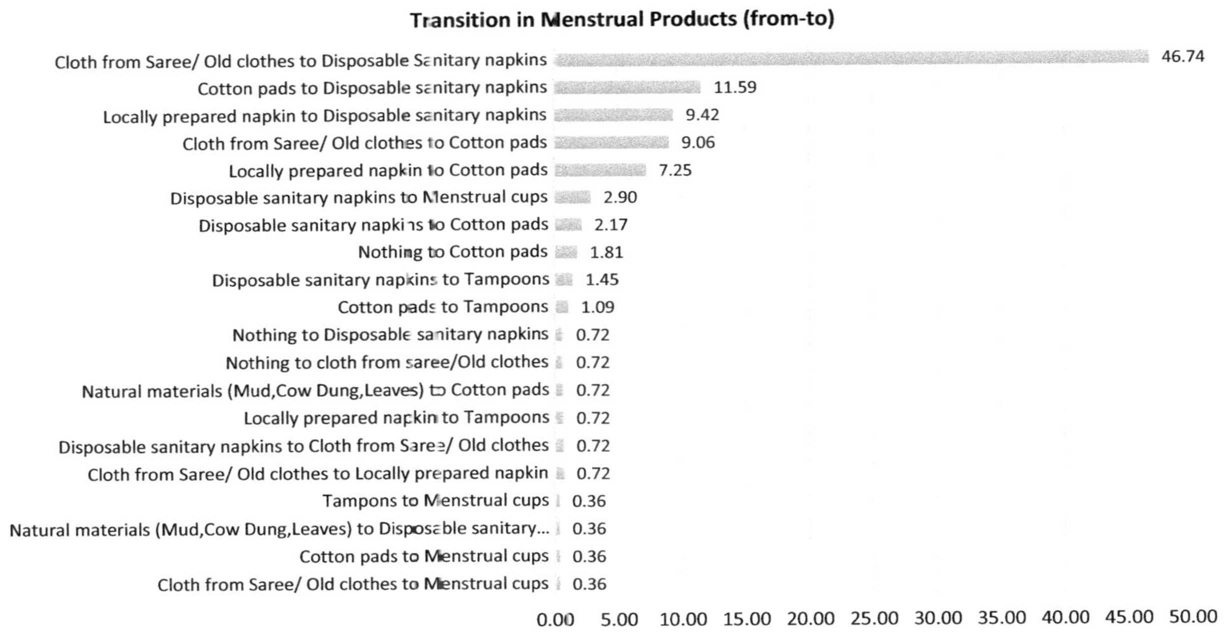

Figure 3.6 Product Preference as Captured by Transition in Menstrual Product.

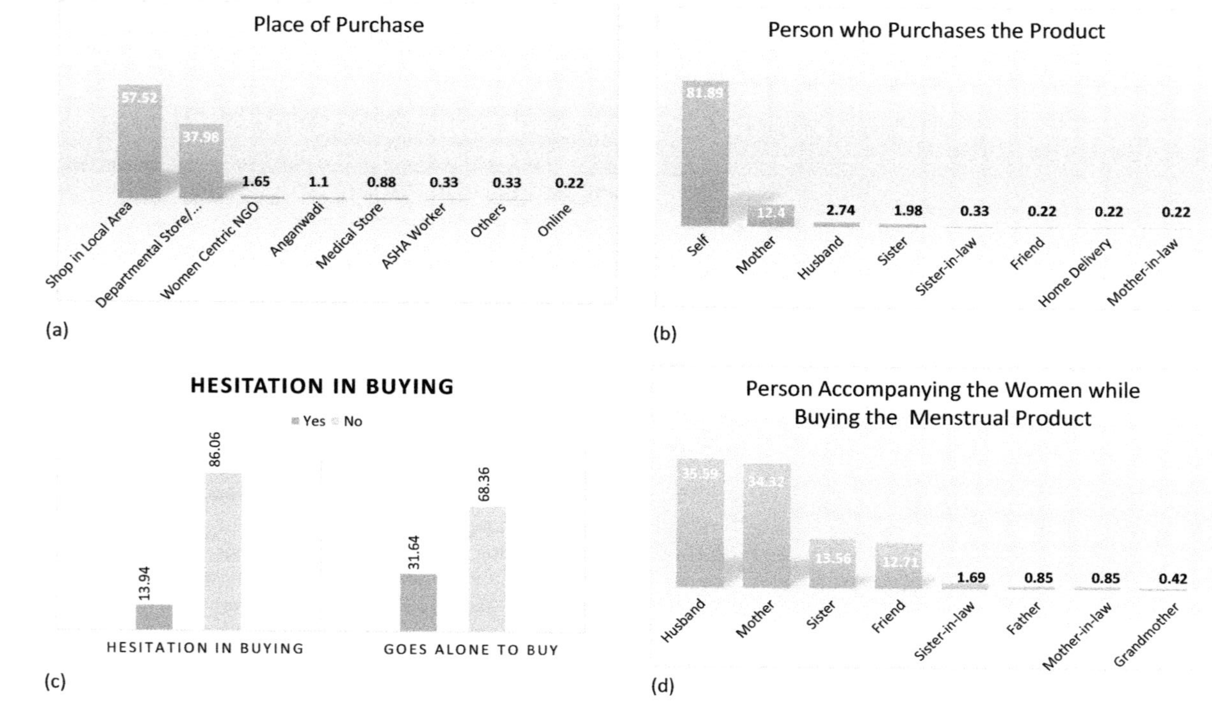

Figure 3.7 Menstrual Product Buying Patterns.

women prefer to buy the menstrual product from the departmental store. A departmental store displays products on shelves and has a separate compartment for menstrual and women-related products. Therefore, those women who buy the menstrual products from the departmental store, get to choose from a variety of products and brands, as compared to those buying from smaller, local stores. However, those buying the menstrual products from Anganwadi, ASHA workers, women-centric NGOs, or other sources have much lesser alternatives to choose from, sometimes the only choice they are required to make is: "to buy or not to buy?". Those purchasing online get even more alternatives to choose from, but online purchases are confined to those who have access to smartphone/tablet/computer, internet, the area is serviced by the online supplier. Also, online purchase of menstrual products requires a change in mindset; at present, most women prefer to buy them offline from stores.

Figure 3.7(b) shows that 81.89 per cent of women buy the menstrual product by themselves, 17.89 per cent get their menstrual product purchased by their friends, husbands, mothers, mothers-in-law, sisters, and sisters-in-law. As seen in Figure 3.7(a), 0.22 per cent purchase their products online, and therefore, they get home delivery of their menstrual products.

Figure 3.7(c) shows that out of 746 women who buy their menstrual products on their own, 86.06 per cent do not hesitate while buying the product. Irrespective of whether they hesitate or not, out of these 746 women, 68.36 per cent women go alone to buy the menstrual product. Therefore, 236 women out of 746 go with a friend, mother, mother-in-law, or some other accompanying person to buy their menstrual products. Figure 3.7(d) shows that out of 236, most married women go with their husbands (35.59 per cent) and most single women go with their mothers (34.32 per cent) to buy their menstrual products. On one hand, there is a taboo on discussing menstruation openly in most societies; two girls in the sample for this study have mentioned going with their respective fathers to buy menstrual products. This could be interpreted as an indication of society gradually opening up and being sensitive towards menstruating women, or that the financial control is in the hands of their fathers.

Figure 3.8 shows that while the family members and husbands of a majority women (65.17 per cent and 71.37 per cent respectively) do not have a say in the choice of menstrual product, 34.83 per cent of women's choice is influenced by their family members Out of the 503 married women, 28.63 per cent women's choices of menstrual products are influenced by their husbands.

The choice of the product not only influences the direct cost but also influences the implicit costs in terms of medical attention needed to treat the side effects of certain products. A discussion on implicit cost is provided in Chapter 4, whereas that of direct costs of menstrual products is discussed in the ensuing section.

Figure 3.8 Family Members'/Husband's Say in the Choice of Menstrual Product.

3.3 Direct Cost of Menstrual Products

Insights into types of costs are important for policy implications. This section describes the method of estimating direct cost of menstrual products for one menstrual cycle. An estimate of usage of disposable sanitary napkins and tampons is also given for a woman for the entire year. The structured questionnaire used to collect data on menstrual hygiene among women of Gujarat seeks information from the respondents about their daily usage of sanitary napkins or tampons for a menstrual cycle. A typical menstrual cycle is six days long and the total of the number of sanitary napkins/tampons used each day, gives an estimate of the usage of sanitary napkins/tampons per menstrual cycle. The respondents are also asked to provide the details of the price of the packet and number of sanitary napkins/tampons contained in that packet. Dividing the price of the packet by the total number of sanitary napkins/tampons in the packet gives an estimate of the rate per piece of a disposable sanitary napkin/tampon. This is then multiplied by the number of sanitary napkins/tampons used during the entire cycle, which gives the cost estimate of menstrual hygiene for one cycle. Table 3.2 provides the summary statistics on the usage and direct cost of disposable sanitary napkins/tampons:

The last column of Table 3.2 gives the number of non-responses for each of these details sought from the respondents. However, in case of usage for each day, it might indicate that the respondent is using multiple products and is not using either disposable sanitary napkin/tampon on that particular day. The respondents who have mentioned cloth from old sarees, cotton pads, natural materials like dung, mud, soil, and menstrual cups are excluded while calculating the summary statistics presented in

Table 3.2 Summary Statistics of Usage and Direct Cost of Disposable Sanitary Napkins/Tampons

Usage/Price	Min.	1st Qu.	Median	Mean	3rd Qu.	Max.	NA's
Day 1 usage	0	2	2	2	3	22	55
Day 2 usage	1	2	2	2	3	21	57
Day 3 usage	0	1	2	2	2	6	62
Day 4 usage	0	1	1	1	2	8	105
Day 5 usage	0	1	1	1	2	1	177
Day 6 usage	0	0	0	0	1	12	296
Disposable sanitary napkins/ tampons used per cycle	0	6	9	9	12	45	47
Price per packet (₹)	6.00	40.00	87.00	134.60	180.00	600.00	37
Disposable sanitary Napkins/ tampons per packet	2	7	10	19	30	240	42
Price per disposable sanitary napkin/tampon (₹)	1.00	5.00	6.67	7.67	10.00	33.33	42
Cost per menstrual cycle (₹)	0.00	37.36	56.00	75.38	91.72	499.95	51
Annual direct cost of menstrual products (₹)	0.00	448.00	672.00	904.50	1101.00	5999.00	51

Note: Sample size of sanitary napkins and tampons users is 607. Therefore, these estimates are based on 607 respondents, whose normally used product is either disposable sanitary napkin or a tampon, not on all the 1025 respondents.

Table 3.2. The respondents who are using cotton pads are excluded because there are two types of disposable cotton pads and one type of reusable cotton pad. Moreover, these cotton pads are also available as raw as well as medicated cotton. Thus, estimating the cost of using cotton pad required more detailing, and as mentioned earlier, we have not attempted estimating the imputed costs. Some respondents had mentioned the cost of cotton pads whereas some did not. As a result, it was difficult to differentiate between the types of the cotton pads used by different respondents. Following up with all the cotton pad users was not possible because of time constraints, and therefore, the estimation was not possible.

The first six rows of Table 3.2 show the summary statistics of the usage of number of disposable sanitary napkins/tampons for each day of the menstrual cycle. The seventh row shows the summary statistics obtained after taking sum of the number of disposable sanitary napkins/tampons for each day of the menstrual cycle, for all six days. Therefore, the zero that appears in the minimum sanitary napkins/tampons used during the entire cycle is actually a very small number and rounds off to zero even if a precision of three places of decimals is taken. The average use of sanitary napkins/tampons per cycle is 9 (both mean and median values). The price of packet of disposable sanitary napkins ranges from ₹6.00 to ₹600.00 with a minimum of 2 pieces per packet to a maximum of 240 pieces per packet. A packet containing 240 is actually a combo of six packs, called the supersaver packs, preferred by women from families having more women in the menstruating age group. Also, sanitary napkins do not have an expiry date, and the expiry date of tampons is 5 years from the date of manufacture. Hence, some women may prefer buying in bulk to make the purchase economical. The packet contains an average of around 19 (≈20) sanitary napkins/tampons (the median is 10).

The price per sanitary napkin/tampon therefore, ranges from as less as ₹1 to as high as ₹33.33, the average price is ₹7.67 (median ₹6.67). Thus, the maximum cost of menstrual products per cycle is ₹499.95; the average cost is ₹75.38, and the median cost is ₹56. On an average, a girl/woman spends ₹904.50 on menstrual products for one year.

Menstrual cups are durable products and can be used for several years before disposing of them. A menstrual cup is made from silicone and has a flexible structure that can be folded to ease the insertion into the vagina, it opens in vagina and collects the menstrual blood; it's a cup. The cup is required to be removed at regular intervals and emptied. The empty cup is washed and reinserted in the vagina to collect more blood. This is an environmentally friendly product, because it is reusable and is relatively easily degradable compared to the material used in disposable sanitary napkins and tampons. Therefore, women are gradually experimenting with menstrual cups, though they are not as popular in India as they are in the western world. In a sample of 1025, there are only twelve menstrual cup users. Therefore, the summary statistics generated cannot be said to be representative of the menstrual cup users but can be used to have a rough cost estimate. Our survey data shows that the minimum price of a menstrual cup is ₹250 and the maximum price is ₹500. The average price of a menstrual cup is ₹392.50 (the median is ₹400). The respondents were not clear on the number of years, a menstrual cup could be used, but van Eijk et al. (2019) show that the average life of a menstrual cup is 10 years. Thus, the annual cost of using a menstrual cup ranges from ₹25 to ₹50, the average cost would therefore be ₹39.25 (the median cost would be ₹40).

This discussion is important from the policy implementation perspective. Scotland is the first country to have made access to menstrual products, free of cost (Lifestyle Desk, 2020). Also, appropriate provisions in the budget for menstrual products could have a pivotal impact on the improvement of public health.

However, usage of menstrual products is only just one, yet the major, component of MHM. There are other direct cost components, and they are discussed in subsequent sections.

3.4 Costs of Allied Menstrual Hygiene Products

Special care is required to be taken during menstrual cycles to ensure the cleanliness of private parts and to keep them disinfected. In order to keep it disinfected, it is required to be rinsed thoroughly either with water or to be cleaned with a soap or a disinfectant like Dettol, or an intimate hygiene product like VWash. If a soap, disinfectant, or an intimate hygiene product is used to clean the private parts, it is a direct cost of menstrual hygiene.

It is also required to thoroughly clean the hands with a soap/hand wash after using the washroom during menstruation. This adds to the direct cost of menstrual hygiene.

This section discusses the estimation of the costs of allied menstrual hygiene products.

Figure 3.9(a) shows the pattern in use of the products for cleaning the private parts, and Figure 3.9(b) shows that pattern of products used for cleaning hands after using the washroom. While a majority (63.51 per cent) of respondents clean their private parts with only water, 68.78 per cent use hand wash to clean their hands after using the washroom. Even today, some women (0.29 per cent) use mud/sand to clean their hands after using the washroom during their menstrual cycles, and only 10.73 per cent of respondents use only water to clean their hands after using the washroom. We sought information on price of the product, and the number of months the product lasts, from those who use either a bathing soap, a disinfectant/antiseptic or an intimate hygiene product to clean the private parts (36.49 per cent respondents) and those who use either a bathing soap, a disinfectant/antiseptic/hand sanitiser or a hand wash (88.98 per cent) to clean hands. This gives an estimate of per-month cost of the product used for cleaning the private parts. However, a soap, disinfectant/antiseptic or an intimate hygiene product is shared by other women members in the menstruating age group in the family, and hence the per-month cost is apportioned by the number of women in the family. Similarly, the soap, hand wash, or the disinfectant/antiseptic/hand sanitiser are largely shared by all the family members in a household. Therefore, the per-month cost of the handwashing products is estimated and apportioned by the number

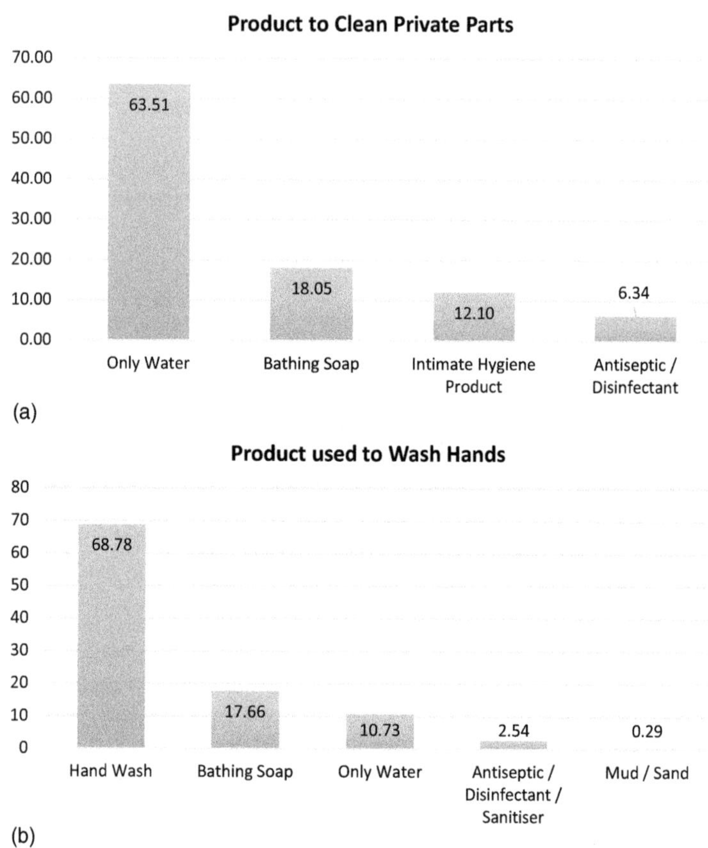

Figure 3.9 Distribution of Products Used for Cleaning Private Parts and Hands after Using Washroom.

of family members in the respondent's household. The summary statistics are then estimated and presented in Table 3.3:

One can see from Table 3.3 that the price of the products used to clean private parts ranges from ₹2 to ₹1000, and these products last for as less as a week to as much as about 3.5 years. The average price of these products is ₹135.10 (median price is ₹100) and these products last, on an average, for slightly more than 2 months and 3 weeks (median: 2 months). The monthly cost of these products, therefore, ranges from ₹2 to ₹500. The average per-month cost of menstrual hygiene products is ₹62.61 (median: ₹50). This cost apportioned by the number of female family members, who might share these products for cleaning their private parts, ranges from ₹1 to

Table 3.3 Estimation of Costs of Allied Products for Menstrual Hygiene Management

	Min.	1st Qu.	Median	Mean	3rd Qu.	Max.	NA's
Cost estimation of the product to clean private parts							
Price of the product to clean private parts (₹)	2.00	50.00	100.00	135.10	180.00	1000.00	218
How long does product to clean private parts last (months)?	0.25	1.00	2.00	2.80	3.00	43.50	215
Per month cost to clean private parts (₹)	2.00	30.00	50.00	62.61	80.00	500.00	221
Per month cost to clean private parts, apportioned by female family members (₹)	1.00	10.00	17.50	30.75	37.50	500.00	221
Annual cost to clean private parts (₹)	12.00	120.00	210.00	369.00	450.00	6000.00	221
Cost estimation of the product to wash hands							
Price of the product to wash hands (₹)	4.00	45.00	72.00	87.62	100.00	380.00	357
How long does the product to wash hands last? (months)	0.250	1.000	1.000	1.456	2.000	9.000	354
Per month cost of the product used to wash hands (₹)	4.50	40.00	57.00	70.85	99.00	380.00	371
Per month cost to wash hands, apportioned by total family members (₹)	1.00	8.00	12.00	15.69	20.00	100.00	371
Annual cost to wash hands (₹)	12.00	96.00	144.00	188.30	240.00	1200.00	371

Note: We have excluded 651 respondents who use only water to clean their private parts and 110 respondents who use only water to wash their hands, for this estimation. It is not based on all 1025 respondents.

₹500, the average is ₹30.75 (median: ₹17.50). Thus, the annual cost ranges from ₹12 to ₹6000, the average annual cost of products used to clean the private parts is ₹369 (median: ₹210).

It may be noted that out of 376 respondents who use products like bathing soaps, disinfectants/antiseptics, intimate hygiene product to clean their private parts, only 156 have responded to the price of the product used by them, and 159 have responded to the number of months, the product lasts. The paired responses, therefore, work out to 153, and the estimation given in Table 3.3 is based on these 153 responses who have given information on both – the price as well as on how long the product lasts.

Table 3.3 also shows that the price of the products used to wash hands (after using the washroom during menstrual cycles) ranges from ₹4 to ₹380 and these products last for as less as a week to as much as nine months. The average price of these products is ₹87.62 (median price is ₹72) and these products last for slightly less than 45 days, on an average (median: 1 month). The monthly cost of these products, therefore, ranges from ₹4.50 to ₹380. The average per-month cost of menstrual hygiene products is ₹70.85 (median: ₹57). This cost apportioned by the total number of family members (male and female, since these products are not menstruation/women-specific), who might share these products for washing hands, ranges from ₹1 to ₹100, the average is ₹15.69 (median: ₹12). Thus, the annual cost ranges from ₹12 to ₹1200, the average annual cost of products used to wash hands after using washroom during menstruation is ₹188 (median: ₹144). At this juncture, we would like the readers to know that we have excluded those cases for cost estimation that have not given details on the price of the product or the time for which the product lasts. We have also excluded those cases who use only water to clean private parts (and similarly, for cleaning hands).

It may be noted that out of 912 respondents who use products like bathing soaps, disinfectants/antiseptics, and hand sanitisers to wash their hands, 555 have responded to the price of the product used by them, and 558 have responded to the number of months the product lasts. The paired responses, therefore, work out to 541 and the estimation given in Table 3.3 is based on these 541 responses, who have given information on both – the price as well as on how long the product, used for washing hands, lasts.

These are going prices at the time of the survey, which was largely during the pandemic in 2020. On one hand, these prices could have been slightly inflated because of the scarcity of products during and soon after lockdown; these are still the market prices during the said period. Therefore, the estimates are based on the market prices. This survey is four years old and might not be reflective of the market prices, when this book is being written. This being said, we could have converted these prices to nominal prices, but for policy prescriptions, market prices better represent the situation. While we do not have the updated information, we do present the formulae for estimating the direct costs of MHM (Appendix – F), which can be used to get updated estimates, even using a different methodology of collecting the data. Policy prescription, relevant to pricing the products, discussed in Chapter 5, is based on these estimates.

3.5 Medical Conditions and Menstrual Dysfunction

Four respondents out of 1025 have mentioned that they are aware that they have PCOS, one respondent has thyroid, one has mentioned persistent pain in lower abdomen, and one has mentioned persistent infection in private parts. While many respondents are aware of the medical conditions they have, only six of them have specified the exact medical condition. PCOS and thyroid influence menstrual cycles. Both PCOS and thyroid can make periods irregular, heavy, or light. There could be a complete absence of periods for a prolonged duration in both circumstances. These women require regular medical treatment to regulate their medical condition and thereby the regularity of menstrual cycles. Thus, the medical costs for women with such medical conditions far exceed the ones who do not have any of these conditions. The prevalence of thyroid among young women is one in eight (Velayutham et al., 2015), and that of PCOS ranges from 3.7 per cent to 22.5 per cent (Ganie et al., 2019) in India.

One of the respondents in the in-depth interview revealed that she got her menarche at the age of 14 years. After her first menstrual cycle, she did not get her next cycle for the next six months. In the beginning, the elderly women family members explained to her that initial cycles could be abnormal, so there is no harm in waiting for the next cycle for a couple of months. However, after waiting for six months, her mother took her to a doctor, and she was treated for regularising her menstrual cycles. She was not diagnosed with any medical condition, and yet she had irregular and erratic menstrual cycles for as long as ten years. This had a deep psychological impact on her, and she had to seek psychiatric advice. The condition of irregular and erratic menstrual cycles is known as menstrual dysfunction, and that can occur even if the woman does not have any medical condition. The mismatch in oestrogen and progesterone levels results in menstrual dysfunction. Not only does this add to the medical cost for its treatment, but it could also have far-reaching implications in requiring one to seek psychological treatment. Thus, medical conditions and menstrual dysfunction can add up to the direct costs of menstruation by several dimensions.

3.6 Summary

Choice of menstrual products is crucial to menstrual hygiene. However, choice is associated with costs, and a rational woman would ideally, aim at optimising her choice (and the resultant utility), subject to budget constraints. Choices are however, constrained by available information and access to the products. This chapter describes the pattern in awareness among women about the different menstrual products available in the market. There is a discussion on the choice of primary product /normally used product and also the choice of secondary product in this chapter. The socioeconomic factors influencing the choice of the product are also discussed. The preference of the outlet to buy the product, whether she buys the product herself or someone else buys it for her, whether she hesitates in buying the product, and whether someone accompanies her while going to buy the product and whether the family members/husband (in case of married women) have a say in the choice of product are probed, and patterns are elicited. Then a discussion on the methodology of direct costs of menstrual products, followed by that of costs of allied menstrual hygiene products, is presented, and the respective cost estimates are worked out. The findings of this study are:

A majority of the respondents are aware about cotton pads, disposable sanitary napkins, and cloth from old sarees. Though there is awareness about tampons and menstrual cups too. The respondents are also aware about the sanitary napkins made available by the government at subsidised rates.

The usage pattern shows that the disposable sanitary napkin is the most preferred absorbent for those using only one product. On examining the influence of socioeconomic characteristics on choice of absorbent, using a binary logistic regression, it is revealed that:

Middle-income groups (both – lower and upper) prefer high hygiene value products compared to the low and high-income group respondents. Disposable sanitary napkins, tampons, and menstrual cups are categorised as high hygiene value products; all others as low hygiene value products.

All the working women and school-going girls prefer high hygiene value products, though college-going girls apparently seem not to.

The region (rural/urban) does not influence the choice of the product. Though women from tribal districts are less likely to prefer high hygiene value products.

Out of 1025 respondents, 32 per cent use more than one product during their menstrual cycles. The most preferred product for those women who use more than one product during their menstrual cycles is cotton pads, followed by disposable sanitary napkins.

Out of 1025 respondents, 410 respondents have changed brands for their primary products, and out of 330 respondents who use more than one product, 147 have changed the brands of their secondary products. Also, out of 1025 respondents, 279 have even switched over from one menstrual product to another. The major reasons for change in brand are identified as improved quality and increased comfort for the primary product. Improved hygiene and an increase in comfort have been the top two reasons (in the given order) for the change in product. A majority of the respondents have switched over from cloth from old sarees to disposable sanitary napkins. A shift from locally prepared napkins to disposable sanitary napkins and cotton pads is also worth noting.

A majority of the women prefer to buy the menstrual product from a local store near their place of residence. This could be because in small towns and rural areas, there might be no departmental store/superstore. The second most preferred place of purchase is departmental store, and that could be a representation of the most preferred alternative in bigger towns. A majority of women themselves go to purchase the menstrual product, though there are a few, for whom these products are purchased by their friends/family members. Absence of a departmental store/superstore and those who do not go on their own to buy the product have constrained choices because of limited amount of information on available products in the market. There are very few (0.22 per cent) who buy online. However, one is required to make a planned purchase while using the online platform because there could be a time gap of a day or two in placing the order and getting the product. The data on hesitation in buying the product shows that the women are now not hesitating in buying the product. While a majority of the women go alone to buy the menstrual product, there are 236 of them who go with their husband/mother to buy the product. While a majority of women are able to make independent choices about which menstrual product to use, 34.83 per cent women's choices are influenced by their family members, and out of 503 married women, 28.63 per cent women's husbands have their say in the choice of menstrual product.

The average annual explicit cost of using disposable sanitary napkins/tampons and that for menstrual cups is given in Table 3.4. The cost of allied products, which is obtained by summing up the average annual cost of using disinfectant/antiseptic/intimate hygiene product/bathing soap for cleaning the private parts and that of using disinfectant/antiseptic/hand wash/hand sanitiser/soap for washing hands, and therefore, it remains the same, irrespective of the type of product used for MHM.

The annual average explicit cost of using disposable sanitary napkins/tampons works out to ₹1461.80 whereas that of using a menstrual cup works out to ₹596.55. Thus, the cost of using a menstrual cup is almost 23 times less than that of using disposable sanitary napkins/tampons, and the total direct cost for menstrual cup users is almost one-third of those

Table 3.4 Average Annual Explicit Costs of Disposable Sanitary Napkins/ Tampons, and Menstrual Cups

Product	Average Direct Cost of Menstrual Hygiene Products (₹)	Average Direct Cost of Allied Products of MHM (₹)	Total Direct Cost of MHM (₹)
Disposable sanitary napkins/ tampons	904.50	557.30	1461.80
Menstrual cups	39.25	557.30	596.55

using disposable sanitary napkins. As mentioned earlier, we haven't estimated the imputed costs for cloths from sarees, cotton pads, or natural materials.

An important direct cost component is because of the frequent medical advice required to be sought for medical conditions like PCOS, thyroid, persistent vaginal infections, and other conditions like endometriosis. All these result in irregular menstrual cycles with too little to too heavy blood flow. These women have to seek treatment on a continuous basis, adding to the medical costs manifold. Women with menstrual dysfunction also require long-term treatment. Women with medical conditions and menstrual dysfunction also experience stress and psychological impact because of the sociocultural beliefs associated with marriage and childbearing. Therefore, these women also have to undergo psychological support/advice, further adding to their medical costs per year.

Notes

1 Anganwadi means "a courtyard shelter" and are government facilities that work primarily towards addressing the nutritional needs of women and children.
2 ASHA, an acronym for Accredited Social Health Activists, are community female health workers appointed by the government. ASHA literally means "hope".
3 "Toxic shock syndrome is a rare but serious medical condition caused by a bacterial infection. It is caused when the bacterium called *Staphylococcus aureus* gets into the bloodstream and produces toxins". (Higuera, 2018).

References

Bae, J., Kwon, H., & Kim, J. (2018). Safety evaluation of absorbent hygiene pads: A review on assessment framework and test methods. *Sustainability, 10*(4146), 1–17. http://dx.doi.org/10.3390/su10114146

Earls, A. O., Axford, I. P., & Braybrook, J. H. (2003). Gas chromatography–mass spectrometry determination of the migration of phthalate plasticisers from polyvinyl chloride toys and childcare articles. *Journal of Chromatography A*, *983*(1), 237–246. https://doi.org/10.1016/S0021-9673(02)01736-3

Fehr, E., & Schmidt, K. M. (2006). Chapter 8 The economics of fairness, reciprocity and altruism – experimental evidence and new theories. In S.-C. Kolm & J. M. Ythier (Eds.), *Handbook of the economics of giving, altruism and reciprocity* (Vol. 1, pp. 615–691). Elsevier. https://doi.org/10.1016/S1574-0714(06)01008-6

Ganie, M. A., Vasudevan, V., Wani, I. A., Baba, M. S., Arif, T., & Rashid, A. (2019). Epidemiology, pathogenesis, genetics & management of polycystic ovary syndrome in India. *Indian Journal of Medical Research*, *150*(4), 333–344.

Higuera, V. (2018). Toxic shock syndrome. *Healthline*, September 17. https://www.healthline.com/health/toxic-shock-syndrome

InformedHealth.org [Internet]. (2017). *Heavy periods: Overview*. Institute for Quality and Efficiency in Health Care (IQWiG). https://www.ncbi.nlm.nih.gov/books/NBK279294/

Lifestyle Desk. (2020). Scotland becomes first country in the world to make sanitary products free; Can India follow suit? *The Indian Express*. https://indianexpress.com/article/lifestyle/health/scotland-first-country-sanitary-products-free-india-7065074/

Santos-Longhurst, A. (2020). How often should you change your pad? *Healthline*, December 15. https://www.healthline.com/health/menstruation/how-often-should-you-change-your-pad

Velayutham, K., Selvan, S. S. A., & Unnikrishnan, A. (2015). Prevalence of thyroid dysfunction among young females in a South Indian population. *Indian Journal of Endocrinology and Metabolism*, *19*(6), 781–784.

van Eijk, A. M., Zulaika, G., Lenchner, M., Mason, L., Sivakami, M., Nyothach, E., Unger, H., Laserson, K. F., & Phillips-Howard, P. A. (2019). Menstrual cup use, leakage, acceptability, safety, and availability: A systematic review and meta-analysis. *The Lancet*, *4*(8), E376–E393. https://doi.org/10.1016/S2468-2667(19)30111-2

Woeller, K. E., & Hochwalt, A. E. (2015). Safety assessment of sanitary pads with a polymeric foam absorbent core. *Regulatory Toxicology and Pharmacology*, *73*(1), 419–424. https://doi.org/10.1016/j.yrtph.2015.07.028

4 Indirect Costs of Menstrual Health and Implicit Costs of Menstrual Hygiene

Menstruation is much more than just 4–6 days of bleeding, resulting from shedding of the uterus lining. During the menstrual cycle, the uterus contracts to ease the shedding of its inner lining. The contraction is aided by the release of prostaglandins, hormone-like substances, that trigger these uterus muscle contractions (Marshburn & Hurst, 2011). This results in cramps/pain. Moreover, changes in levels of oestrogen and progesterone would result in mood swings, anxiety and irritability; and serotonin affects mood, emotions and thoughts, and 90 per cent of menstruating women experience pre-menstrual syndrome (PMS) (Higuera, 2019). It is recommended that in such situations, women should consult a doctor if the symptoms persistently worsen, or remain constant but disrupts the routine, even for a day or two. On the other hand, women appear to be quite naïve about what causes menstruation, menstrual cramps, and other ailments associated with menstruation. In our study, we do not attempt to examine the medical causes of menstrual cramps, or any other issues experienced by a woman during her menstrual cycles or a couple of days prior to its onset, commonly known as the pre-menstrual syndrome or just PMS.

There are direct implications of these ailments/syndromes on medical expenses required to be incurred to treat these. This would include consultation charges of the doctor if one avails the treatment in a private healthcare facility and the costs of medicines. The implications of these syndromes/ailments are: (1) adverse impact on a woman's quality of life, even for a day or two, yet every month, before/while she is having her menstrual cycles. (2) school absenteeism (and thereby, on studies): the implications of missing school are difficult to quantify. Vashisht et al. (2018) observe that in Delhi (India), 40 per cent of girls miss school and 65 per cent of them miss routine school activities because of PMS, menstrual ailments, and lack of access to absorbent. (3) Workplace absenteeism: the

DOI: 10.4324/9781003473961-4

implications of missing work would result in wage loss to the women, especially those, who work as casual labourers or even in the unorganised sector.

Sociocultural taboos restrict women from doing household chores during their monthly menstrual cycle. While in most households, these chores are done by other female family members, or in some households, by male and female family members jointly, or by neighbours, there are very few households that may require to employ paid services for these household chores. This constitutes one of the indirect cost components of menstruation.

This chapter is organised into six sections: Section 4.1 discusses the prevalence of PMS and ailments experienced by women during menstruation. It provides a methodology for estimating the medical costs of PMS and menstruation ailments and provides a yearly cost estimate for the same. Section 4.2 shows patterns in the reasons for absenteeism, the percentage of women who remain absent for a day or two during their menstrual cycles, and the methodology of estimating the yearly opportunity cost of absenteeism for working women. Section 4.3 describes the ailments in private parts because of failure to comply with menstrual hygiene and estimation of their yearly medical costs. Section 4.4 describes the estimation of yearly costs of doing household chores like cooking, cleaning etc., by employing a paid help, apportioned to the days of the menstrual cycle. The issues in estimating these costs are also discussed at length. This chapter focuses on examining the indirect costs of menstrual health and implicit costs of menstrual hygiene, required to be borne by the woman. However, there is a brief mention of the potential environmental costs because of inappropriate disposal (partly because of lack of awareness, and partly because of lack of facilities) of used menstrual product waste in Section 4.5. Finally, Section 4.6 summarises the discussion on indirect and implicit costs.

The basic objective of this chapter is to identify the indirect and implicit cost components of menstruation, to describe the methodology of estimating the respective costs, and wherever possible, provide a yearly estimation of these cost components. In-depth interviews were conducted to examine in detail the different problems faced by women with PMS, during their monthly menstrual cycle, as a result of some medical condition, or simply because of menstrual dysfunction; the interview guide can be found in Appendix C. The narratives of these in-depth interviews are provided in appropriate sections, with the objective of providing a deeper understanding of the indirect/implicit costs of menstruation faced by women.

Based on PPI and consultation with gynaecologists and GPs dealing with reproductive health issues (interview guide can be found in Appendix A), we have identified broad indirect and implicit cost components of menstruation as shown in Figure 4.1.

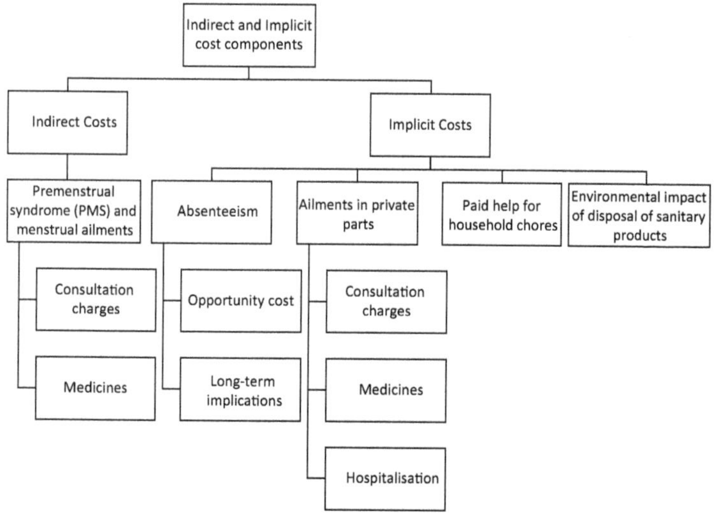

Figure 4.1 Indirect Cost Components of Menstrual Health and Implicit Cost Components of Menstruation.

It may be noted that while most costs are borne by the individual, the medical costs of treatment would be borne by the state in the case of state-funded clinics and hospitals. Also, the implicit costs for paid leaves are borne by the employer, and environmental costs are borne by the society.

4.1 Pre-Menstrual Syndrome, Menstruation Ailments and Medical Attention

A large number of women experience psychological and physical ailments during their pre-menstrual days as well as during their menstrual cycles. Zita (1988) highlights depression, irritability, sleep disturbances, lethargy, alcoholic excesses, vertigo, nausea, vomiting, constipation, bloating, enuresis, urinary retention, migraine headaches, increased susceptibility to infection, suicidal tendencies (or attempts), work morbidity, manic reactions, and dermatological diseases as PMS. In-depth interviews with gynaecologists and GPs dealing with reproductive health issues were conducted for this study. They reveal two broad categories of ailments that are experienced by women, either a few days, typically a week before the onset of the menstrual cycle every month and/or even during the days while she is menstruating: psychological symptoms and physical symptoms. A list of psychological and physical ailments was prepared and then discussed with women of different ages and socioeconomic strata

(PPI work), before finalising the questionnaire for this study. Taking Zita (1988) as the base, subsequent discussion with the gynaecologists and doctors practicing women's health issues, and PPI activities involving discussion with women and culturally contextualising them to Gujarat (India), the following symptoms have been identified and examined for this study:

Psychological symptoms: anxiety, confusion, forgetfulness, irritability, mood swings, anger fits, difficulty concentrating, and depression.

Physical symptoms: headache, stomachache, muscle fatigue, general fatigue, dizziness, giddiness, sweating/palpitation, nausea, vomiting, diarrhoea, fainting, breast tenderness, and bloating.

The respondents were asked to select, whether they experienced each of these symptoms prior to the beginning of the menstrual cycle, that is, as PMS, during menstrual cycles, both – as PMS as well as during the menstrual cycles, or they don't experience it at all. The results are shown in Figures 4.2 and 4.3:

Among the psychological problems, anxiety is the most common of all PMS; 24.88 per cent of women experience anxiety as PMS. All other PMS are experienced by 11–20 per cent of women in the sample. During menstruation, anger fits, mood swings, and irritability are experienced by 37.07, 36.20 and 32.78 per cent of women, respectively, whereas 28.98 per cent of them experience difficulty in concentrating during their menstrual cycles. However, mood swings are experienced by 17.37 per cent of women, both as PMS and during their menstruation cycle. Irritability and anger fits are experienced by 16.49 and 16.20 per cent of women, both as PMS and during menstruation. Other symptoms are found in 3–9 per cent of women, both as PMS and continue to exhibit even during their menstrual cycles.

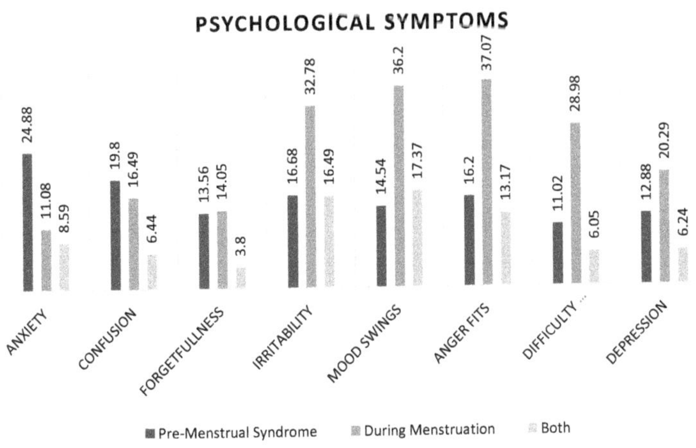

Figure 4.2 Psychological Symptoms Prior to Menstruation/during Menstruation.

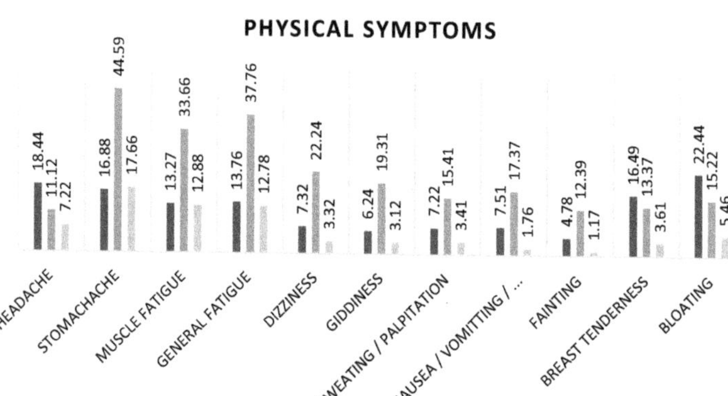

Figure 4.3 Physical Symptoms Prior to Menstruation/during Menstruation.

The commonest of physical problems is bloating, experienced by 22.44 per cent of women, followed by headache, experienced by 18.44 per cent of women, and stomachache, experienced by 16.88 per cent of women as the PMS. Other physical problems are found to be experienced by 4–14 per cent of women in their pre-menstrual phase. The commonest of physical ailments experienced during the menstrual cycles are stomachache (44.59 per cent), general fatigue (37.76 per cent) and muscle fatigue (33.66 per cent). Other physical ailments are experienced by 11–22 per cent of women, while in their menstrual cycles. Also, 17.66 per cent of women experience stomachache, both, as PMS and during their menstrual cycles, and close to 13 per cent of women experience general fatigue and muscle fatigue, both as PMS and during menstruation. All other physical ailments are experienced by 1 to 7 per cent of women, both as PMS and extending through their menstrual cycles.

Theano (1968) shows that one-third of the girls/women who experience PMS or ailments during menstruation have to seek medical advice. A majority of the women (6 out of 7) with whom we did an in-depth interview, for this study also mentioned requiring medicines to manage their PMS and menstruation ailments. The results of the in-depth interview showed that the women are hesitant, and most of them, do not even consider it necessary to consult a doctor for managing the psychological symptoms. This could be because of the cultural taboos associated with exhibiting one's psychological disturbances, even if they are associated with the release of hormones for their monthly cycles. Women also revealed during the in-depth interviews that they prefer over-the-counter pain relievers or home remedies for their physical ailments. However, only in one case, the woman had to go to the doctor to get a pain-reliever injection. Discussing the details about her stomachache, she told us that she would tell her mother as soon as she experienced menstrual cramps. Then her mother would tell her to go to the nearest Primary Health Centre (PHC) and tell the doctor to give her a pain-reliever injection. Neither the doctor asked the reason for her asking for a pain-reliever injection, nor did the mother guide her to discuss this with the doctor. This continued for several years until she got married, and her ailment gradually reduced after childbirth. This is because discussing anything pertaining to menstruation, including menstrual cramps, was a taboo, around 20 years back. Even today, women do not go to the doctor for stomachache, headache, muscle fatigue, or general fatigue. However, in case of extreme pain or unbearable symptoms, they seek medical advice. In this study, out of 1025 respondents on whom questionnaires were administered, 69 women (and that is 6.74 per cent) have sought medical attention for various PMS and menstrual ailments. Based on the details given by them, we have attempted to estimate the cost of PMS/menstrual ailments. However, before moving on to cost estimates, we would offer a discussion on the methodology to arrive at those estimates.

We asked our survey respondents, the number of times in the past year, they sought medical advice. This gives the number of consultation spells in the past year. The respondents were asked about the number of days, they required to undergo treatment in each spell. If the respondents had more than one spell of seeking medical advice, they were asked to mention the average days per spell. Then, they were asked about the number of times they were required to consult the doctor in each spell. For example, if the treatment per spell was spread over, say, 5 days, they were asked about the number of times they were required to see the doctor during these 5 days. Again, if there were more than one spell of seeking medical advice, the average number of times they required a doctor's consultation per spell was asked. Finally, we asked them about the average consultation charges, each time they saw the doctor. The consultation charges per year are then calculated by multiplying the number of spells in the past year with the number of times they were required to consult the doctor, and this figure is then multiplied by the consultation charges per check-up.

Similarly, the costs of medicines in past year are calculated by multiplying the number of spells with the average cost of medicines per spell.

The formulae for estimating the yearly costs of consultation charges and medicines can be found in Appendix G (Equations 3 and 4).

The total medical costs of treating PMS/menstruation ailments per year are then calculated by adding up the total consultation charges per year and total medicine costs per year. The range of costs associated with pre-menstrual syndrome/menstrual ailments is given in Table 4.1.

Table 4.1 Estimation of Costs of PMS/Menstruation Ailments

Cost Components	Min.	1st Qu.	Median	Mean	3rd Qu.	Max.	NA's
Number of times, consulted a doctor in past year	1.000	1.000	1.500	2.706	3.000	25.000	1
Number of days received treatment per spell	1.00	2.00	6.00	19.30	25.50	180.00	6
Frequency of consultation per spell	1.000	1.000	1.000	1.873	2.000	8.000	6
Consultation charges per checkup (Rs)	0.00	100.00	200.00	252.30	400.00	1000.00	0
Consultation costs per spell (Rs)	0.00	100.00	250.00	485.60	600.00	5000.00	4
Medicines costs per spell (Rs)	0.00	100.00	300.00	594.00	600.00	5000.00	11
Consultation costs per year (Rs)	0.00	150.00	400.00	1811.00	1000.00	25000.00	12
Medicines cost per year (Rs)	0.00	185.00	650.00	1883.00	1050.00	50000.00	12
Total cost of PMS/menstruation ailments per year (Rs)	0.00	400.00	1100.00	3604.00	2105.00	68750.00	12

Note: The costs of PMS/menstrual ailments are estimated at Market Prices and at the unit level (separately, for each respondent), and then summary measures are generated for each category of costs. Therefore, the aggregated cost figures for different summary measures in Table 4.1 might not correspond to the figures worked out using the formulae of Appendix G. In order to reproduce the figures of Table 4.1, you will need access to the respondent-level data.

Table 4.1 shows that women are required to consult the doctor up to a maximum of 25 times a year. It might be intriguing that if a woman menstruates 12 times a year, how is it possible that she consults a doctor for more than twice the frequency of menstruation? The answer is: some women informed us during the in-depth interview that they were undergoing prolonged treatment for as long as six months to several years. However, there have been breaks in between the treatments, and the question is asked about the number of spells in the past year only. It is likely that her menstrual cycle is more frequent than the average 28-day span. There are two such outliers in this sample: one who has sought medical advice for 15 spells in past year, and the other one who has sought it for 25 days. All other women have sought medical advice up to 11 times in the past year of the survey. The average spells of seeking medical advice are close to 3 times (median: 2 times, rounded off to the nearest integer) in the past year.

It may be noted that the reference period for this cost estimation is the past year from the survey date. Since the survey was undertaken during the pandemic, the past year is referred to as normal circumstances. Therefore, these costs give an estimate of normal circumstances and not the pandemic time. During the pandemic, it would have been much lower. This is because, during the pandemic, all doctors focused on treating COVID-19 patients. Moreover, people did not seek medical consultations unless it was absolutely necessary, or it was a part of their ongoing routine checkups/treatments.

The number of days per spell ranges from 1 day to 180 days. Some women are required to take a prolonged treatment for 6 months on a continuous basis, and therefore, it is likely that she might have had only one spell in the past year but would have lasted for six months. On average, the number of days per spell is 19 (median: 6 days) for which they are required to undergo treatment. The frequency of consultation per spell ranges from one to eight times, an average of two days per spell (median: once per spell).

Maximum one-time consultation charges per checkup are ₹1000; some women have not been required to pay consultation charges, especially if they are visiting a government healthcare facility or if the doctor is closely known to them. The average one-time consultation charges are ₹252.30 (median: ₹200). This multiplied by the number of times a woman consulted the doctor per spell, estimates the details of consultation charges per spell. The maximum consultation charges per spell are ₹5000; average charges are ₹485.60 (median: ₹250).

The consultation charges per spell multiplied by the number of spells in the past year give the consultation charges in the past year. The summary statistics of consultation charges in the past year reveal that the maximum was ₹25000, and the average was ₹1811 (median: ₹400).

The maximum cost of medicines per spell as per our survey data is ₹5000. Some respondents have mentioned even receiving medicines for free. This is because they could be undergoing treatment in some trust-based (charity) clinics, or the doctor was known to them and was not prescribing the medicines but giving them from their clinic. The average cost of medicines per spell is ₹594 (median: ₹300). Therefore, the total costs of medicines per year would be the number of spells in the past year multiplied by the cost of medicines per spell. The summary statistics of medicine costs per year show that the maximum cost is ₹50000, and the average cost is ₹1883 (median: ₹650).

The total burden of treatment for PMS and menstrual ailments, thus, can go up to a maximum of ₹68750 and an average of ₹3604 (median: ₹1100). This is one of the two indirect cost components of menstruation.

Considering that there are vast variations in the spell of treatment, consultation charges, and cost of medicines, we have used median values to even out those variations. We would, however, recommend using these formulas separately for the type of medical services (government, private, or charity) to arrive at more accurate estimates.

4.2 Absenteeism

Several studies have shown that girls remain absent from school (Al Mutairi & Jahan, 2021; Boosey et al., 2014; Chaghlana et al., 2019; Grant et al., 2013; Oster & Thornton, 2011; van Eijk et al., 2016) during their menstrual cycles, largely because of taboos associated with attending schools during menstrual cycles. Our study explores the reasons for school/work absenteeism during menstrual cycles.

Herrmann and Rockoff (2012) observe that there is no evidence of a gender gap in absenteeism at work, implying no menstruation-related absenteeism. Absenteeism from work could be paid or unpaid; most organisations have a system of giving a limited number of paid leaves to employees during the year. However, in a majority of cases, the absences are unpaid. This is more common among casual labourers and employees of the unorganised sector. The statistics on absenteeism, based on this study, are given in Figure 4.4:

Figure 4.4(a) shows that 26.05 per cent of women stay absent from their school/work. The number of absent days, typically ranges between one to two days, when the blood flow is heavy. Chaghlana et al. (2019) observe in their study that girls do not go to school out of fear of blood leakage and staining their outer clothes.

The reasons for absenteeism are examined and can be seen in Figure 4.4(c). In this study, the major reason (73.41 per cent) for absenteeism is that the woman feels uncomfortable and tired during the days of heavy blood flow. The next most cited reason is the fear of staining the outer

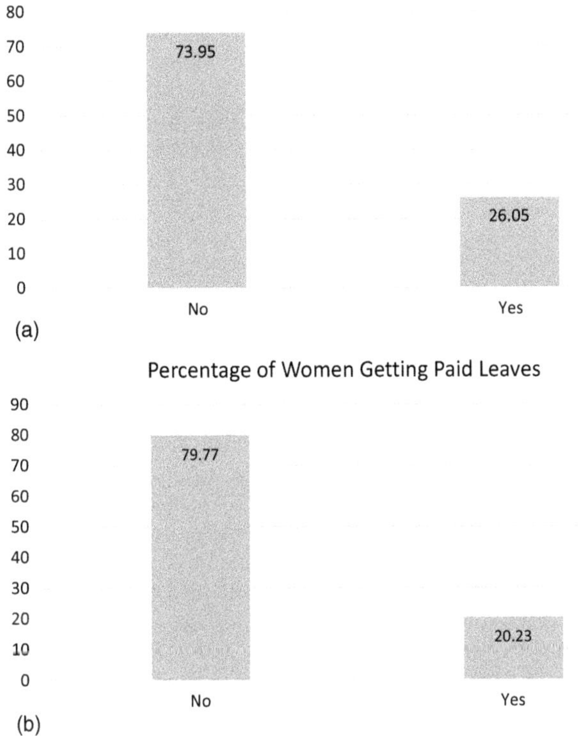

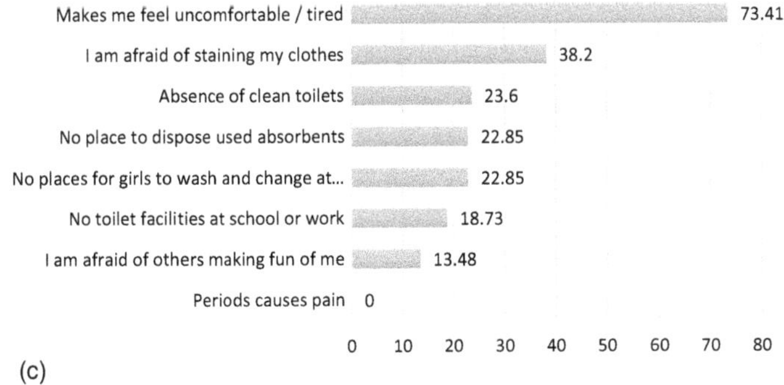

Figure 4.4 Absenteeism due to Menstruation and the Opportunity Cost.

clothes (38.20 per cent). Babagoli et al. (2022) have found that the use of sanitary napkins intervention has a cost-effectiveness of $300 per student-year in reducing absenteeism in schools, whereas that of menstrual cups is $2300 per disability-adjusted life years (DALY) averted. These observations are helpful from a policy perspective.

Our study observes that there are other reasons for absenteeism from school/work, like the absence of clean toilet facilities at work (23.60 per cent), no place to dispose of the used sanitary napkins and no place to change the sanitary napkins/cloths (22.85 per cent each), absence of toilet facilities (18.73 per cent) and fear of others making fun (13.48 per cent). It is important to note that not a single woman remains absent because of menstrual pain. However, across cultures, the major cause of absenteeism during menstrual cycles is menstrual pain, plus menstrual hygiene management issues like difficulties in changing absorbents, fear of staining and smelling, and absence of WASH facilities and absence of disposal facilities (Chaghlana et al., 2019; Kumbeni et al., 2021; Miiro et al., 2018; Mohammed et al., 2020; Shah et al., 2022; Vashisht et al., 2018).

This issue was probed further through the in-depth interviews. All the women interviewed were working women. These women whom we interviewed were sampled from both unorganised sector as well as from the organised sector. However, irrespective of the nature of their employment and their entitlement to paid leaves, they mentioned that they did not remain absent from work because of period pain. They rather managed by taking an over-the-counter pain reliever and still attended to their duties. Our study suggests that 173 out of 512 total working women in our sample[1], remain absent for a day or two during their menstrual cycles. Out of these 173 women, 79.77 per cent of women do not get paid leaves (Figure 4.4(b)).

The implicit cost of menstruation for a woman would be calculated as the number of days she remains absent from work (which is normally one or two per month) multiplied by her wages/salary per day. The total number of absent days because of menstruation per year multiplied by her average per-day wages/salary would give the estimate of the yearly implicit cost of menstruation for the working woman.

On the other hand, for the women who get paid leaves, the number of leaves due to menstruation multiplied by her per-day wages/salary is the opportunity cost to the employer per woman employee of the organisation. Opportunity cost estimates to an employer are complex and based on a lot of assumptions, with short-term and long-term productivity loss (Navarro & Bass, 2006; Pauly et al., 2002), which explains the dearth of literature on estimates of implicit/opportunity costs of illness, let apart menstrual issues. However, the objective of this study is to examine/estimate the cost of menstruation to women, and therefore, a detailed discussion on opportunity cost to employers is not much attempted here.

4.3 Menstrual Hygiene and Ailments in Private Parts

There could be several reasons why a woman would knowingly ignore menstrual hygiene. Some of the reasons could be cultural: not being able to dry the reusable cloth in open space (preferably in the sun), observing seclusion during the days of monthly cycles, not being allowed to take a bath during the menstrual cycles, and others (as discussed in Chapter 2). Some could be circumstantial: working women who might not have adequate facilities in the workplace to change the sanitary napkins/tampons at regular/stipulated intervals. These could result in the neglect of menstrual hygiene. Studies by Brabin et al. (1998), Durr-e-Nayab, (2005), and Mulgaonkar (1996) show that neglect of hygiene during menstruation can result in reproductive tract infections (RTI). Inappropriate use of tampons/being hard on pushing the tampons has resulted in vaginal rashes (Cardwell, 1942). The RCTs of Chow and Bartlett (1989) show that tampons alter the autochthonous vaginal microflora. There is no consensus on whether toxic shock syndrome (TSS) is caused by tampons (Vostral, 2017). Moreover, TSS could be fatal, and therefore, we have not included this in our study. Costing of fatal infections and terminal illnesses requires a different approach to costing.

On the other hand, in-depth interviews with gynaecologists and GPs dealing with women's reproductive problems reveal that itching in private parts, vaginal infection, rashes in/around the vaginal tract, moniliasis (a type of fungal infection), and abscess could be because of the negligence in hygiene practices during menstruation. According to them, herpes and internal swelling resulting in pain in the pelvic region are not direct results of neglected hygiene. Itching in private parts and vaginal infection are the two most common ailments observed among women. Figure 4.5 shows that 23.41 per cent of women have experienced itching in private parts, this is followed by rashes in/around the vaginal tract, which is experienced by 9.56 per cent of women. Women who experienced other ailments range between 1 per cent and 4 per cent.

Both the gynaecologists and the respondents of in-depth interviews have discussed that the sanitary napkins with higher absorbing capacities, which apparently add to the comfort and keep the private parts dry and the foul smell away, can cause itching and vaginal infections. The polymers used in the high-absorbing-capacity sanitary napkins could be harsh on the tender skin of the private parts. One respondent, while narrating her experience about the use of disposable sanitary napkins, revealed that she used cloth from old sarees for a very long time, as she belonged to a small town and there was hardly any exposure to alternative mensuration products when she attained menarche. After she moved to a larger town and had access to television, she learnt about the disposable sanitary napkins. On the recommendation of her friends in the new urban

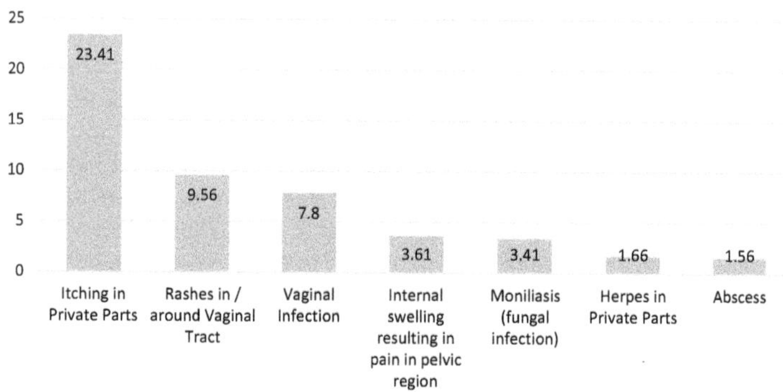

Figure 4.5 Percentage of Women who have Suffered from Different Ailments in their Private Parts.

surroundings, she ventured to buy and started to use the sanitary napkins. However, with prolonged use, she started developing rashes. Though, initially, she could not make out that the rashes are due to the use of disposable sanitary napkins. She would apply soframycin ointment, and in a few days, she would start feeling better. However, once she realised that the itching has continued and it is deep down, rather than superficial. Applying soframycin was not possible. Hesitantly, she went to the doctor, only to learn that the material of the disposable sanitary napkin was too harsh on her skin. She continued to change products, but being a working woman and requiring staying out of her home for at least eight hours a day, she switched back to disposable sanitary napkins and must take medicines every month after her menstrual cycles to get the infection treated. She did not attempt using a menstrual cup and did not disclose the reasons for not attempting to use it. A similar incident was narrated by another woman who too had realised that she was allergic to disposable sanitary napkins. She mentioned that along with the disposable sanitary napkins, which she carries in her office bag as a preparedness for the monthly cycle, she also always keeps antifungal medicines that are recommended by her doctor.

All these have a bearing on the monthly expenditure on menstruation and thus, constitute opportunity costs of menstrual hygiene, some of which arise out of non-hygienic practices but also as a side effect of hygienic absorbents. The method of calculating the financial burden of seeking medical advice for these ailments is the same as described for the PMS and menstruation ailments in Section 4.1. However, very few respondents have given information about the number of spells, consultation charges per checkup, number of consultations per spell, consultation charges, medicine costs, etc., for ailments in private parts. Most respondents told us that they don't remember. Therefore, it was not possible to estimate the opportunity costs associated with these ailments.

Unlike PMS and menstrual ailments, these ailments in private parts, especially abscesses or severe and recurrent herpes, could lead to hospitalisation, and in extreme situations, might require surgery. The respondents have mentioned the number of hospitalisation episodes from one to five times in their entire menstruation span. Moreover, we haven't included menopausal women. Women in the survey are at different stages in their reproductive age span. Therefore, this estimate of hospitalisation is not accurate and would require more probings/details in terms of hospitalisation until a particular age. While we have information on respondents' age, we do not have detailed information on number of hospitalisations from menarche till a particular age. Moreover, they informed us that they are not able to recall the expenditure incurred during the hospitalisation episode. All this was too much for them to recall. Therefore, a different methodology is required to be worked out to estimate these costs.

Thus, opportunity cost of menstrual hygiene, be it cultural or circumstantial, could be exorbitant, as this would range from treating the ailments, to hospitalisation and even surgery, depending upon the nature of ailment.

4.4 The Cost of Household Chores

It was discussed in Chapter 2, Section 2.3, that very few women practice the taboo of doing household chores and cooking. This is contrary to most studies, which show that the women are required to observe cooking restrictions (Srinivasa & Manasa, 2017; House et al., 2012; Mahon & Fernandes, 2010) during their menstrual cycles, though there is no mention of restrictions on other household chores in these studies. Figure 4.6(a) and (b) respectively for the percentage of women who observe taboos, and the percentage of women who hire paid help. The results of our study on restrictions on doing various household chores, as shown in Figure 4.6(a), reveal that 64.59 per cent of women have no restrictions on cooking during menstruation, and more than 70 per cent of women have no restrictions on cleaning utensils, washing clothes, and fetching water.

For the remaining (30 per cent or less) women who observe restrictions on cleaning utensils, washing clothes, and fetching water during menstruation, these tasks are performed either by other female family members of the household, who are not having the menstrual cycles at the time. Alternatively, these are performed by any family member of the household (man or woman), or they are just asked to stay away, irrespective of whoever does this. In some communities, the neighbours and relatives also

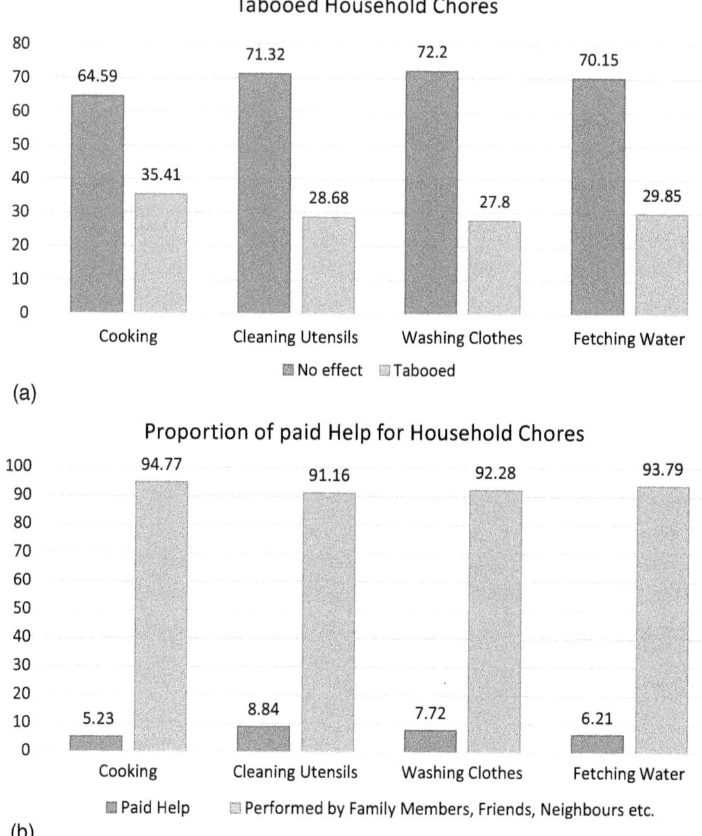

Figure 4.6 Effect of Menstruation on Household Chores.

help each other with household chores while the woman is in her menstrual cycle. However, as shown in Figure 4.6(b),

- Out of 35.41 per cent of respondents who have restrictions on cooking, 5.23 per cent hire paid help for cooking (or they buy food from restaurants/food outlets),
- Out of 28.68 per cent who have restrictions on cleaning utensils, 8.84 per cent hire paid help,
- Out of 27.80 per cent who have restrictions on washing clothes, 7.72 per cent hire paid help, and
- Out of 29.85 per cent women who have restrictions in fetching water, 6.21 per cent hire paid help.

The implicit cost of household chores is applicable to those women who observe restrictions and hire paid help for getting these chores performed (including buying food from outside). Normally, the restriction on household chores, including cooking, is observed for 3–5 days from the onset of each menstrual cycle. For each household chore, for example, cooking, the number of days per month when cooking is tabooed, multiplied by the per-day wages to be paid to the cook or per-day expenditure on buying cooked food from a restaurant/food outlet gives an estimate of the per-month cost of cooking during the menstrual cycle. This per-month estimate is then multiplied by 12 to get the yearly estimate of the cost of cooking during menstruation. The yearly cost of cleaning utensils, washing clothes, and fetching water is calculated in the same manner. All these costs are added up to get the annual estimate of the cost of household chores during menstrual cycles. The summary statistics of each of these annual costs and the total annual cost of household chores would give an idea about the minimum, maximum, and average values of the respective annual costs of household chores during menstrual cycles.

However, the range of the wages of a cook and domestic help for other chores varies in great proportions across the geographical regions of Gujarat, also between rural and urban areas, and also across developed, developing, and tribal districts. The wages of a cook and domestic help would also depend on the number of family members in a household. Since the majority of respondents could not answer the details of the wages of the paid domestic help, it was not possible to estimate the cost figures of household chores.

This issue was probed during the in-depth interviews too. However, most of the respondents of the in-depth interview mentioned not having any restrictions or their family members taking over these chores while she is having her monthly menstrual cycles. In-depth interviews also provided insights about the domestic help for cleaning utensils and washing clothes. In a vast majority of the urban households and a few well-to-do rural households, they employ domestic help for cleaning utensils and washing clothes. These domestic helps are hired on fixed monthly wages. There is a trend of domestic help working for a long number of years with the same family. These domestic helps go an extra mile to do other household chores like fetching water or cooking, *albeit*, on payment of an agreed lump sum amount. This makes it difficult for the respondents to give an estimate about the paid help per month, and that too, exclusively for her monthly menstrual cycle.

On the other hand, in an in-depth interview, who worked as a cook for a few families, revealed that she does not observe any taboos at her own home. However, she has to inform one/two of her employers about her menstrual cycles and stay absent for three days since the onset of her cycle, every month. Thus, even if she is a paid help, hired on fixed monthly wages, she has to observe restrictions on cooking during her monthly menstrual cycle at her employer's kitchen. Nevertheless, she is being paid for those number of days; she remains absent during her menstrual cycles.

4.5 Potential Environment Impact of Disposal of Menstrual Products Waste

Proper disposal of used menstrual products is essential to have a minimum impact on the environment, as all the menstrual products are not completely biodegradable. Most sanitary napkins contain polymers and other chemicals that are not completely biodegradable (Foster & Montgomery, 2021). Therefore, throwing the sanitary napkins/tampons in routine waste or burning them would be harmful to the environment. Alzate and Sánchez (2020) estimate that per menstrual cycle, per woman, 832 grams of solid waste is generated in a year, working out to 8.32 kg in 10 years, which is the life span of a menstrual cup. Flushing the sanitary napkins/tampons through the sewer can choke the sewer line in long run, causing undesirable effects of issues with disposal of wastewater. The most appropriate method could be disposing of it in an incinerator. Yet, most developing countries lack a system of disposal of sanitary napkins/tampons, and therefore, a meagre proportion of women have access to incinerators. Also, there is little awareness among the users about the appropriate method of disposal. The details on the method of disposal of sanitary napkins/tampons provide evidence that 12.03 per cent of women are putting the sanitary napkins/tampons through the incinerator (Figure 4.7).

This study does not make any attempt to estimate the environmental costs, but Figure 4.7 provides insights into the potential impact on the

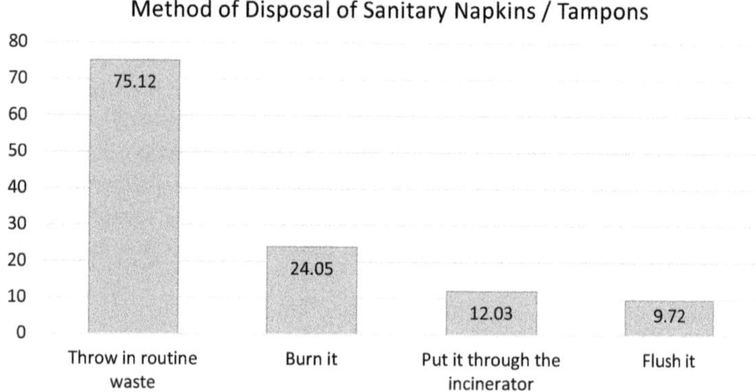

Figure 4.7 Method of Disposal of Sanitary Napkins/Tampons.

environment, that could be caused by inappropriate disposal. It may be noted that some women dispose of the sanitary napkins through more than one method, and hence, the total of percentages in Figure 4.7 adds up to more than 100 per cent.

Menstrual cups are relatively easily biodegradable. However, in our study, only 12 users of menstrual cups could be traced. Out of them, only seven have disposed of their menstrual cup so far. This is because menstrual cups were a relatively new concept in India when this survey was undertaken, and these cups have a long lifespan. Out of these seven women who have got to dispose of their menstrual cups, four of them have disposed of it as routine waste, whereas three have made efforts to dispose of it at its recycling facility.

4.6 Summary

Menstrual cycles have much more than just bleeding for four to six days each month. Ninety per cent of women experience PMS (Chumpalova et al., 2020), some ailments like stomach ache, headache, etc., continue for a couple of days even after the onset of monthly menstrual cycles. In some girls/women, these symptoms are so severe that they are required to remain absent from school or work. It is observed in many other studies, including this, that girls/women stay absent from school or work for fear of staining their outer clothes. Thus, the cost of menstruation far exceeds the costs of using the menstrual products and the cost of products used to ensure the hygiene of private parts and for washing hands. The costs required to be incurred for PMS or menstruation-related ailments, are indirect costs, and the costs that arise because of not being able to adhere to menstrual hygiene are implicit/opportunity costs. This chapter presented a discussion on the indirect and implicit cost components of menstruation, the methodology to estimate these, and wherever possible, providing an annual estimate for the said cost components. The results of this study are summarised in this section as follows:

The medical costs of PMS and menstrual ailments constitute indirect costs of menstrual health. However, before proceeding to describe the methodology of estimating the medical cost of these ailments, the prevalence of PMS and menstrual ailments among the sample chosen for this study is examined. The list of psychological and physical ailments experienced by a woman a couple of days prior to her menstruation and during her monthly menstrual cycle is prepared based on the past studies and discussion with gynaecologists and GPs dealing with women's reproductive issues. The results of this study show that anxiety is the commonest of all the other PMSs, though anger fits, mood swings, and irritability are more common during the menstruation phase. Among the physical symptoms, bloating, headache, and stomachache are common PMS, whereas stomachache, muscle fatigue, and general fatigue are experienced more during the menstrual cycles' phase. Some women experience these syndromes right from the pre-menstrual days and continue to experience them for the initial 2–3 days of menstrual cycles. While a large proportion of women prefer to take over-the-counter drugs, some women require to consult a doctor in the event of extreme symptoms. The yearly consultation charges of the doctor plus the yearly expenditure on medicine to treat these ailments constitute the medical costs of PMS and menstrual ailments, and it could be as high as ₹68750; the average costs are ₹3604 (median: ₹1100).

The first implicit cost component is the opportunity cost of absenteeism from work. The results show that some women prefer to remain absent for a day or two during the days of heavy blood flow out of fear of staining their outer clothes. The annual opportunity cost of absenteeism because of menstruation is estimated by adding up the total number of menstrual absenteeism per year, multiplied by the average daily wage/per day salary of the woman, who is not entitled to paid leaves. The study focuses on the individual implicit costs of menstruation, but a woman entitled to paid leave will result in an equivalent amount of opportunity cost to the employer. There is difficulty in quantifying the opportunity cost of absenteeism of a school-going student/the one pursuing higher education (not in the workforce and studying). It may be noted that while opportunity cost is a one-time cost, the cost of absenteeism has long-term implications in terms of promotions for the working women and on examination performance and career choice/progression for students.

The second component of implicit cost results from medical advice sought for different ailments like, itching in private parts, infection in private parts, rashes in/around the private parts, and in extreme situations, severe infections like herpes, abscess, moniliasis, and internal swelling in the pelvic region. These problems could be because of neglect in menstrual hygiene or because of the adverse impact of the menstrual products used to soak blood. Improper use of menstrual products, negligence in hygiene because of a lack of adequate and clean sanitation facilities available in school/at the workplace or being allergic to certain chemicals used in menstrual products may result in these infections/itching. The method of estimating the medical costs of these ailments is similar to that of estimating the medical costs of PMS and menstrual ailments. However, in this component of implicit cost, the cost of hospitalisation required to treat abscess or severe moniliasis is also to be added. The study shows that the number of hospitalisation episodes ranges from once a year to five times in the entire span from menarche till this survey was conducted. All the respondents, however, do not recall the expenditure incurred on consultation charges, costs of medicines, and other medical expenses during hospitalisation, and hence, an estimate could not be worked out for this component.

The third component of the implicit cost of menstruation is the annual expenditure on paid services hired for tabooed household chores – cooking, cleaning utensils, washing clothes, and fetching water. The annual cost of household chores is estimated by multiplying the average daily wages of the paid domestic help by the number of days of observing these taboos per month, multiplied by 12. The number of days for observing taboos ranges from three to five. The difficulty in estimating the annual costs of household chores is that some households give paid leaves to the domestic help every month for three to five days while she is having her monthly cycles. This may, at times, overlap with the menstrual cycle of the employer of the paid domestic help, which complicates the calculations.

It has been emphasised in this chapter that this study focuses on individual implicit costs of menstruation. However, the issues associated with the disposal of used menstrual product waste provide insights into the environmental costs, a cost that is borne by society. This requires being addressed in the public health policy, and hence, we have provided a brief discussion on the environmental costs of menstrual waste disposal towards the end of this chapter.

Note

1 Total sample size is 1025 girls/women, out of which 512 are working women.

References

Al Mutairi, H., & Jahan, S. (2021). Knowledge and practice of self-hygiene during menstruation among female adolescent students in Buraidah city. *Journal of Family Medicine and Primary Care, 10*(4). https://journals.lww.com/jfmpc/fulltext/2021/10040/knowledge_and_practice_of_self_hygiene_during.13.aspx

Alzate, E., & Sánchez, Y. (2020). The use of sanitary pads and menstrual cups and their impact on environment. *2020 Congreso Internacional de Innovación y Tendencias En Ingeniería (CONIITI)*, 1–5. https://doi.org/10.1109/CONIITI51147.2020.9240278

Babagoli, M. A., Benshaul-Tolonen, A., Zulaika, G., Nyothach, E., Oduor, C., Obor, D., Mason, L., Kerubo, E., Ngere, I., Laserson, K. F., Tudor Edwards, R., & Phillips-Howard, P. A. (2022). Cost-effectiveness and cost-benefit analyses of providing menstrual cups and sanitary pads to schoolgirls in rural Kenya. *Women's Health Reports (New Rochelle, N. Y.)*, *3*(1), 773–784. https://doi.org/10.1089/whr.2021.0131

Boosey, R., Prestwich, G., & Deave, T. (2014). Menstrual hygiene management amongst schoolgirls in the Rukungiri district of Uganda and the impact on their education: A cross-sectional study. *The Pan African Medical Journal, 19*, 253–253. PubMed. https://doi.org/10.11604/pamj.2014.19.253.5313

Brabin, L., Gogate, A., Karande, A., Khanna, R., Dollimore, N., DeKoning, K., Nicholas, S., & Hart, C. A. (1998). Reproductive tract infections, gynaecological morbidity and HIV seroprevalence among women in Mumbai, India. *Bulletin of the World Health Organization*, *76*(3), 277–287.

Cardwell, M. G. (1942). Tampons in menstruation. *The British Medical Journal*, *1*(4242), 537.

Chaghlana, S. P. S., Amaravadhi, S. R., Mazodi, S. D., & Lokusu, V. S. (2019). Determinants of menstrual hygiene among school-going adolescent girls in urban areas of Hyderabad. *International Journal of Community Medicine and Public Health*, *6*(5), 2211–2215.

Chow, A. W. & Bartlett, Karen H. (1989). Sequential assessment of vaginal microflora in healthy women randomly assigned to tampon or napkin use [with discussion]. *Reviews of Infectious Diseases*, 11, S68–S74. JSTOR.

Chumpalova, P., Iakimova, R., Stoimenova-Popova, M., Aptalidis, D., Pandova, M., Stoyanova, M., & Fountoulakis, K. N. (2020). Prevalence and clinical picture of premenstrual syndrome in females from Bulgaria. *Annals of General Psychiatry*, *19*, 3–3. PubMed. https://doi.org/10.1186/s12991-019-0255-1

Durr-e-Nayab. (2005). Reproductive tract infections among women in Pakistan: An urban case study. *The Pakistan Development Review*, *44*(2), 131–158. JSTOR.

Foster, J., & Montgomery, P. (2021). A study of environmentally friendly menstrual absorbents in the context of social change for adolescent girls in low- and middle-income countries. *International Journal of Environmental Research and Public Health*, *18*(18). https://doi.org/10.3390/ijerph18189766

Grant, M., Lloyd, C., & Mensch, B. (2013). Menstruation and school absenteeism: Evidence from rural Malawi. *Comparative Education Review*, 57(2), 260–284. JSTOR. https://doi.org/10.1086/669121

Herrmann, M. A., & Rockoff, J. E. (2012). Does menstruation explain gender gaps in work absenteeism? *The Journal of Human Resources*, *47*(2), 493–508. JSTOR.

Higuera, V. (2019). PMS (Premenstrual syndrome). *Healthline*, May 11. https://www.healthline.com/health/premenstrual-syndrome#_noHeaderPrefixedContent

House, S., Mahon, T., & Cavill, S. (2012). *Menstrual hygiene matters: A resource for improving menstrual hygiene around the world*. WaterAid.

Kumbeni, M. T., Ziba, F. A., Apenkwa, J., & Otupiri, E. (2021). Prevalence and factors associated with menstruation-related school absenteeism among adolescent girls in rural northern Ghana. *BMC Women's Health*, *21*(1), 279. https://doi.org/10.1186/s12905-021-01418-x

Mahon, T., & Fernandes, M. (2010). Menstrual hygiene in South Asia: A neglected issue for WASH (water, sanitation and hygiene) programmes. *Gender and Development*, *18*(1), 99–113. JSTOR.

Marshburn, P. B., & Hurst, B. S. (2011). *Overview: Disorders of menstruation*. Blackwell Publishing Ltd.

Miiro, G., Rutakumwa, R., Nakiyingi-Miiro, J., Nakuya, K., Musoke, S., Namakula, J., Francis, S., Torondel, B., Gibson, L. J., Ross, D. A., & Weiss, H. A. (2018). Menstrual health and school absenteeism among adolescent girls in Uganda (MENISCUS): A feasibility study. *BMC Women's Health*, *18*(1), 4. https://doi.org/10.1186/s12905-017-0502-z

Mohammed, S., Larsen-Reindorf, R. E., & Awal, I. (2020). Menstrual hygiene management and school absenteeism among adolescents in Ghana: Results from a school-based cross-sectional study in a rural community. *International Journal of Reproductive Medicine*, *2020*(1), 6872491. https://doi.org/10.1155/2020/6872491

Mulgaonkar, V. B. (1996). reproductive health of women in urban slums of Bombay. *Social Change*, *26*(3 & 4), 137–156.

Navarro, C., & Bass, C. (2006). The cost of employee absenteeism. *Compensation & Benefits Review*, *38*(6), 26–30. https://doi.org/10.1177/0886368706295343

Oster, E., & Thornton, R. (2011). Menstruation, sanitary products, and school attendance: Evidence from a randomized evaluation. *American Economic Journal: Applied Economics*, 3(1), 91–100. JSTOR. https://doi.org/10.2307/25760247

Pauly, M. V., Nicholson, S., Xu, J., Polsky, D., Danzon, P. M., Murray, J. F., & Berger, M. L. (2002). A general model of the impact of absenteeism on employers and employees. *Health Economics*, *11*(3), 221–231. https://doi.org/10.1002/hec.648

Shah, V., Nabwera, H., Sonko, B., Bajo, F., Faal, F., Saidykhan, M., Jallow, Y., Keita, O., Schmidt, W.-P., & Torondel, B. (2022). Effects of menstrual health and hygiene on school absenteeism and drop-out among adolescent girls in rural Gambia. *International Journal of Environmental Research and Public Health*, *19*(6). https://doi.org/10.3390/ijerph19063337

Srinivasa, S., & Manasa, G. (2017). Menstrual hygiene among school-going adolescent girls. *TJPRC: International Journal of General Pediatrics and Medicine*, *2*(2), 17–22.

Theano, G. (1968). the prevalence of menstrual symptoms in Spanish students. *British Journal of Psychiatry*, *114*, 771–773.

van Eijk, A. M., Sivakami, M., Bora Thakkar, M., Bauman, A., Laserson, K. F., Coates, S., & Phillips-Howard, P. A. (2016). Menstrual hygiene management among adolescent girls in India: A systematic review and meta-analysis. *BMJ Open*, *6*. http://dx.doi.org/10.1136/bmjopen-2015-010290

Vashisht, A., Pathak, R., Agarwalla, R., Patavegar, B. N., & Panda, M. (2018). School absenteeism during menstruation amongst adolescent girls in Delhi, India. *Journal of Family and Community Medicine*, *25*(3). https://journals.lww.com/jfcm/fulltext/2018/25030/school_absenteeism_during_menstruation_amongst.4.aspx

Vostral, S. (2017). Toxic shock syndrome, tampons and laboratory standard-setting. *CMAJ: Canadian Medical Association Journal = Journal de l'Association Medicale Canadienne*, *189*(20), E726–E728. https://doi.org/10.1503/cmaj.161479

Zita, J. N. (1988). The premenstrual syndrome: "Dis-easing" the female cycle. *Hypatia*, *3*(1), 77–99. JSTOR.

5 Discussion and Policy Prescriptions

The Right of Women to Menstrual Leave and Free Access to Menstrual Health Products Bill was introduced in 2022, though it has not yet been passed by the parliament. The bill proposes to provide for three paid menstrual leaves per month for "women and transwomen" employees. The bill further proposes to ensure free access to menstrual products. The legislation was proposed to be applicable to both the private and public sector (The Right of Women to Menstrual Leave and Free Access to Menstrual Health Products Bill, 2022). In 2017 too, an attempt was made to pass a bill on menstruation benefits (Menstruation Benefits Bill, 2017). While the 2017 bill was not introduced in Lok Sabha,[1] the 2022 bill was introduced but not passed.

There have been debates around providing menstrual leave: the issue of inclusivity if a woman is absent for four (three menstrual leaves and one general leave) days each month. This can also influence the employability of women's subsequent promotion.

The long-term effect of three days a month of menstrual leave could be discriminatory from the social perspective. Also, from an economic perspective, it would have long-term cost implications for the woman herself, and this would also add to the cost of the employer because of reduced productivity by three days each month.

On the other hand, ensuring free menstrual products could have a financial bearing on the government (increase in the costs to the government); it would still be a support to all those women who cannot afford to buy hygienic menstrual products. We have provided a detailed discussion on the opportunity costs of menstrual hygiene in Chapter 4.

Menstruation is inevitable and is experienced by a woman for almost two-thirds of her life span. Despite that, it continues to face the social taboos. The connotations about monthly menstrual cycles subtly implied that women are weaker compared to men, that women experience monthly sickness, or as the days of monthly infirmity of women. This

DOI: 10.4324/9781003473961-5

might have emanated from the fact that absorbents with high blood-absorbing capacity were not available then. It was immaterial whether menstruation implied women's inferiority to men, or vice versa, menstruation did confine the role of women to domestic duties only.

The Ministry of Health and Family Welfare (MoHFW) introduced a scheme[2] in 2011, with the objective of increasing awareness about menstrual hygiene and increasing the access and use of good quality sanitary napkins among adolescent girls (10–19 years age group). The scheme aims to ensure safe and environmentally friendly disposal of used sanitary napkins. This scheme was initially launched in 107 districts in 17 Indian states and is scaled up to 157 districts across 20 Indian states. Under this scheme, a pack of 12 sanitary napkins for ₹12 is made available to rural adolescent girls, through the Accredited Social Health Activists (ASHAs). The ASHA is rewarded with ₹1 for each pack sold, plus she is entitled to one free pack every month. The ASHA is required to convene meetings at the Anganwadi Centres (AWCs) every month to guide the adolescent girls on menstrual hygiene and on sex and reproductive health (SRH). Information, Education and Communication (IEC) material is developed to create awareness about menstrual hygiene and is available in audio, video, and reading material formats.

5.1 Broad Observations from our Study

Age-old beliefs and taboos associated with menstruation have resulted in poor awareness about menstruation among women, and that makes menstrual hygiene management (MHM) difficult. Menstrual health is necessary not only for ensuring good reproductive health but also for ensuring overall good health. We undertook a study in Gujarat (India) to examine the extent of awareness about menstruation and menstrual hygiene among women. We also examined the prevalent beliefs and taboos associated with menstruation. The basic objective of this study was to estimate costs associated with menstrual hygiene, and the opportunity costs of not being able to adhere to hygienic practices during menstruation.

This study extensively quotes the available literature on awareness and practices, as well as on beliefs and taboos in Chapter 2, alongside presenting and discussing the findings from our study about the same. There is a dearth of literature on costs associated with menstrual health and hygiene. This study is an attempt to identify economic cost components of menstrual health and hygiene and provide broad estimates, wherever possible, through the primary survey.

Some literature is available to understand the availability of menstrual products, facilities for change and disposal, and absenteeism during menstrual cycles. There are mixed results on menstruation-associated absenteeism on wage loss and loss in education attendance. While the estimation of opportunity cost associated with absenteeism is possible for working women, it is difficult to estimate the opportunity cost of loss of education because of absenteeism in school/higher education. The inability to adhere to menstrual hygiene can result in costs of treating ailments that would arise from poor hygiene. These too add to the opportunity cost of menstrual hygiene. On the other hand, there are implicit costs associated with medical conditions like thyroid and PCOS, resulting in irregular menstrual cycles. There could be additional costs despite using hygienic products because of their side effects.

Menstrual and pre-menstrual syndromes also result in medical expenditure and constitute the indirect cost of menstruation.

We have attempted to identify direct, indirect, and implicit (opportunity) costs of menstruation, menstrual hygiene, and health.

The survey for our study is conducted in three phases over a period of six months using a mixed methods approach. Though, costs are determined largely through questionnaire data, whereas cost components were elicited through in-depth interviews with doctors and women. The survey work coincided with the first wave of COVID-19, and therefore, major adaptations had to be done for the survey work. The details of the original plan and adaptation have been presented in Chapter 1 (Section 1.3).

We collected data from women living in both rural and urban areas, and covered districts that are developed, developing, and tribal (defined in Chapter 1, Section 1.3) to ensure the representation of women from different socioeconomic strata. We also ensured representation of women across the reproductive age group, education levels, and nature of the occupation. The main findings from our survey are presented below.

5.1.1 *Awareness about Menstruation and MHM*

Mother plays an important role in providing awareness about menstruation and allied issues to the girls at menarche. Even during the initial few menstrual cycles, girls mostly receive guidance and awareness from their mothers. Compared to awareness about physical changes that take place at puberty, there is relatively less awareness about hormonal changes that take place around menarche/at puberty.

Awareness seems to increase with level of education. Awareness is found to be less among non-working women as compared to working women or school/college-going girls. Women from urban areas have higher awareness than those from rural areas. The awareness among women from developed districts is higher compared to those from developing and tribal districts. Thus, the government should devise programmes to focus more on rural areas and ensure the involvement of non-working women in awareness programmes.

5.1.2 Beliefs about Menstruation

A large proportion of respondents do not believe that irregularity in menstrual cycles is a disease. They also don't believe that regular menstrual cycles with a frequency of either less than a month or more than a month are a disease.

Educated women have milder beliefs. Non-working women seem to have stronger beliefs. Women, especially in urban areas and in developed districts, have stronger beliefs surrounding menstruation. This requires the awareness programme to focus more on myth eradication for non-working women in urban areas and developed districts.

5.1.3 Taboos Practiced during Menstrual Cycles

While taboos on religious activities are strictly practiced even today by a large proportion of respondents, the taboos associated with household activities like doing household work, cooking, touching stored food, and sleeping on the usual bed seem to have gotten milder with time and are practiced only by a few respondents. However, a majority of the respondents don't seem to be carrying on a routine life during menstruation. This could be because of the menstrual cramps, pain, or mood swings during the menstrual cycles.

A lesser number of taboos are practiced with an increase in educational attainment. The practices of taboos by the women in urban areas are milder compared to those living in rural areas; those living in developing or tribal districts have milder taboos. The practices of taboos by women in joint and extended families are milder compared to those from nuclear families. Christians, Parsis, and Sikhs observe milder taboos compared to Hindus, Muslims, and Jains; Sridhar (2019) observes that Sikhs do not observe any taboos about menstruation. Thus, awareness programmes for taboo eradication should be targeted in rural areas and developed districts. Awareness programmes may be targeted to enhance awareness among religious leaders, which can in turn be passed on to the society (women and their families).

5.1.4 Usage Pattern of Sanitary Products

A vast majority of the respondents are aware only of cotton pads, disposable sanitary napkins, and a piece of cloth torn from old sarees. A relatively few have awareness about tampons and menstrual cups. The respondents are also aware of the sanitary napkins made available by the government at subsidised rates.

The usage pattern shows that disposable sanitary napkins are the most preferred absorbent for those using only one product. The influence of socioeconomic characteristics on the choice of absorbent is examined using a binary logistic regression. Middle-income groups (both – lower and upper) prefer high hygiene value products compared to the low and high-income group respondents; the definition of income groups is also discussed in Chapter 1 (Section 1.4). Disposable sanitary napkins, tampons, and menstrual cups are categorised as high hygiene value products, all others as low hygiene value products. The results further reveal that:

- There is a direct relation between level of education and preference for high-value mensuration products.
- All the working women and school-going girls prefer high hygiene value products, as compared to college-going girls.
- The region (rural/urban) or the extent of the development status of the district (developed/developing/tribal) does not influence the choice of the mensuration product.

Out of 1025 respondents, 32 per cent use more than one product during their menstrual cycles. The most preferred product by women, who use more than one product during their menstrual cycles, is cotton pads, followed by disposable sanitary napkins. There could be many reasons for using more than one product.

Women have changed brands and products over time. The major reasons for the change in the brand are identified as improved quality and improved comfort for the primary product. The topmost reason for the change in brand of the secondary product is 'improved quality' which is followed by 'environmentally friendliness' of the product. Improved hygiene and increase in comfort have been the top two reasons (in the given order) for change in product. A majority of women in our study have switched over from a piece of cloth from an old saree to disposable sanitary napkins. A large number of women have also switched over from locally prepared napkins to disposable sanitary napkins and from cotton pads to disposable sanitary napkins.

Most women prefer to buy menstrual products from a local store in the vicinity of their place of residence. This could be because, in small towns and rural areas, there might be no departmental store/superstore. The second most preferred place of purchase is departmental store, and that could be a representation of the most preferred alternative in bigger towns/cities. Majority of women themselves go to purchase the menstrual product, though there are a few for whom these products are purchased by their friends/family members. Choices are constrained for those who do not go on their own to buy menstrual products. Cities/towns that do not have departmental stores/superstores also pose a constraint on the choice of menstrual products, even if the woman buys the product by herself. There are very few women who buy products online.

The data on hesitation in buying the product shows that women are now not hesitating in buying menstrual products. While a majority of the women go alone to buy the menstrual product, there are some women/girls whose husbands/mothers accompany them to buy the product. Also, a majority of women are able to make independent choices about which menstrual products to use. Yet, some women's choices about menstrual products are influenced by their family members (and husbands, in the case of married women).

5.1.5 Costs Associated with Menstruation, Menstrual Health and Hygiene

We have estimated the average annual cost of using disposable sanitary napkins/tampons and that for menstrual cups separately. Total direct costs are obtained by adding this to the average yearly cost of using a product to clean and disinfect the private parts and that used for washing hands. The annual average direct cost of using disposable sanitary napkins/tampons works out to be ₹1461.80, whereas that of using a menstrual cup works out to be ₹596.55. These estimates are inclusive of the allied products of menstrual hygiene. Thus, the cost of using a menstrual cup is almost one-third of using disposable sanitary napkins/tampons.

Women with medical conditions like PCOS, endometriosis, and thyroid have to bear additional expenses each month for menstrual health. These conditions result in irregular menstrual cycles with too little to too heavy blood flow. Women having these conditions have to seek treatment on a continuous basis, adding to the medical costs manifold. Women with menstrual dysfunction also require a long-term treatment. Women with medical conditions and menstrual dysfunction also undergo stress because of the sociocultural beliefs associated with marriage and childbearing. Therefore, these women also have to undergo psychological support/advice, which further adds to the annual direct costs in terms of the costs of medical consultation and prescription medicines.

Menstrual cycles have much more than just bleeding for four to six days, each month. Ninety per cent of women experience PMS (Chumpalova et al., 2020). The results of this study show that anxiety is the commonest of all the other PMSs, though anger fits, mood swings, and irritability are more common during the menstruation phase. Among the physical symptoms, bloating, headache, and stomachache are common PMS, whereas stomachache, muscle fatigue, and general fatigue are experienced more during the menstruation phase. Some women experience these syndromes right from the pre-menstrual days and continue to experience them for the initial 2–3 days of menstrual cycles. While a large proportion of women prefer to take over-the-counter drugs, some women require a consultation with a doctor in the event of extreme pain. The medical cost of PMS and menstrual ailments constitutes indirect costs. The annual consultation charges of the doctor plus the annual expenditure on medicine to treat these ailments constitute the medical costs of PMS and menstrual ailments, and it could be as high as ₹68,750; the average costs are ₹3604 (median: ₹1100).

There are implicit (opportunity) cost components, which are summarised below.

Absenteeism is one of the major implicit cost components of menstruation. The results show that some women prefer to remain absent for a day or two during the days of heavy blood flow, out of fear of staining their outer clothes. The annual opportunity cost of menstrual leaves is estimated by adding up the total menstrual leaves per year, multiplied by the average daily wage/per day salary of the woman, for those who are not entitled to paid leaves. There is difficulty in quantifying the opportunity cost of absenteeism of a school-going student.

The second component of implicit cost results from medical advice sought for different ailments like, itching in private parts, infection in private parts, rashes in/around the private parts, and in extreme situations, severe infections like herpes, abscess, moniliasis, and internal swelling in the pelvic region. These problems could be because of neglect in menstrual hygiene or because of the adverse impact of the menstrual products used to soak menstrual blood. Improper use of menstrual products, negligence in hygiene because of a lack of adequate and clean sanitation facilities available in school/at the workplace or being allergic to certain chemicals used in menstrual products may result in these infections/itching. The method of estimating the medical costs of these ailments is similar to that of estimating the medical costs of PMS and menstrual ailments. However, in this component of implicit cost, the cost of hospitalisation required to treat abscess or severe moniliasis is also to be added. The study shows that the number of hospitalisation episodes ranges from once a year to five times in the entire span from menarche till this survey was conducted.

The third component of implicit cost of menstruation is the annual expenditure on paid services hired for cooking, cleaning utensils, washing clothes, and fetching water. The yearly cost of household chores is estimated by multiplying the average daily wages of the paid domestic help multiplied with the number of taboo days per month, multiplied by 12. The number of taboo days ranges from three to five during the mensuration cycle. However, there are some difficulties in arriving at the precise estimate.

This study focuses on an individual's implicit costs of menstruation. However, the issues associated with the disposal of used menstrual product waste provide insights into the environmental costs, a cost that is borne by society, which has not been attempted in this study.

5.2 The Areas that Require Attention of the Policy-Makers

There are many NGOs operating in the area of spreading awareness and distributing disposable sanitary napkins, either at subsidised rates or for free. Some are working in rural areas, some in local areas, whereas some others have a geographical coverage across districts/states. The coverage could be increased manifold by having a coordinated approach with the NGOs.

The scheme for the promotion of menstrual hygiene among adolescent girls needs to be scaled up to include any menstruating women (between menarche and menopause), rather than only adolescent girls. Also, adolescent girls are defined as those between 10 and 19 years, for the said scheme. However, the average age of menarche among Indian girls is close to 14 years (Pathak et al., 2014). Thus, the scheme effectively caters to girls between 14 and 19 years. A vast majority of the menstruating women are not included in the scheme.

The scheme is launched only in the rural areas under the National Health Mission (NHM). There is a requirement to cater to the urban slums and women from low-income groups living in urban areas. In an in-depth interview with one of the NGOs, the founder informed that she started her services by distributing disposable sanitary napkins in some select schools in one mega city of Gujarat.[3] Soon, she learnt that the girls were not using the sanitary napkins. Instead, they continue using the traditional methods of soaking menstrual blood. Initially, the girls were hesitant to speak, but later on, they broke their silence. The founder of the NGO said that it was shocking to learn that the girls could not afford an underwear. The disposable sanitary napkins require an underwear to glue it to. Therefore, she started distributing the underwear along with the sanitary napkins; as a result, the usage of napkins increased manifold. She observed that the girls who used the napkins informed their mothers about the comfort provided by them, and they induced their mothers to buy the low-cost disposable sanitary napkins. Not only did the girls talk with their mothers, but they also talked with other girls in the vicinity, which resulted in a multiplier effect. A small gesture of distributing sanitary napkins with underwear went a long way to inculcate menstrual hygiene habits among women of all age groups through the beneficiaries of this NGO. The government can provide for making available a low-cost underwear of different sizes to be sold along with the disposable sanitary napkins. While sanitary napkins are a one-time-use product, the same underwear can be used for a longer duration, normally, one year. Products like reusable cotton pads or tampons can be promoted that would not require women to buy underwear.

Reusable cotton pads are environmentally friendly. Similarly, menstrual cups are also environmentally friendly. The yearly direct costs of menstrual hygiene for menstrual cups have worked out to one-third of that for sanitary napkins/tampons. Scotland is the first country that has made the menstrual products absolutely free (Lifestyle Desk, 2020). Reduction in direct costs of menstrual products can help India to make certain products like menstrual cups/reusable cotton pads for free and charge subsidised rates for the disposable sanitary napkins/tampons.

This study examines the extent of the financial burden, in terms of explicit cost of menstrual hygiene. Women from very poor backgrounds find it difficult to wash their hands with a soap/disinfectant/hand sanitiser after changing the sanitary napkin/any other material used to soak menstrual blood. This NGO started distributing kits containing a pack of sanitary napkins, a small soap bar (just to last the five/six days of the monthly menstrual cycle) and a sachet of shampoo to wash their hair. The government can take the initiative to distribute such hygiene kits and keep it voluntary for woman to buy only a pack of sanitary napkins, only soap bars, only shampoo sachet, or a combination of any of these, or all three.

One Gujarat-based NGO has started to hire men as volunteers for the menstrual hygiene activities. Though, further details on the nature of work undertaken by these men volunteers could not be sought, this is certainly a good initiative, as it will create awareness about menstruation among men. The awareness among women is very low, and efforts are required to increase awareness among women. However, if simultaneous efforts are made to create and increase menstruation awareness among men, it would result in a psychologically healthy and matured society.

NGOs can involve women in co-operative and self-help groups for the distribution or even for the manufacturing of low-cost sanitary napkins. The government may also provide grants to the NGOs for such activities.

Notes

1 The Parliament of India has two houses: Lok Sabha (the lower house) and Rajya Sabha (the upper house). Any bill has to be passed by both the houses in order to be enacted as an Act.
2 https://nhm.gov.in/index1.php?lang=1&level=3&sublinkid=1021& lid=391, accessed on March 28, 2021, accessed again on October 20, 2024 for updates, if any.
3 Names and geographical locations of NGOs are not mentioned to comply with the confidentiality and anonymity clauses of the ethical guidelines of this survey.

References

Chumpalova, P., Iakimova, R., Stoimenova-Popova, M., Aptalidis, D., Pandova, M., Stoyanova, M., & Fountoulakis, K. N. (2020). Prevalence and clinical picture of premenstrual syndrome in females from Bulgaria. *Annals of General Psychiatry, 19*, 3–3. PubMed. https://doi.org/10.1186/ s12991-019-0255-1

Lifestyle Desk. (2020). Scotland becomes first country in the world to make sanitary products free; Can India follow suit? *The Indian Express.* https://indianexpress.com/article/lifestyle/health/scotland-first-country-sanitary-products-free-india-7065074/

Menstruation Benefits Bill, 16, Lok Sabha, Government of India XV (2017). https://eparlib.nic.in/handle/123456789/1465061?view_type=browse

Pathak, P. K., Tripathi, N., & Subramanian, S. V. (2014). Secular Trends in Menarcheal Age in India: Evidence from the indian human development survey. *PloS One*, *9*(11), 1–13. https://doi.org/10.1371/journal.pone.0111027

Sridhar, N. (2019). *Menstruation across cultures: A historical perspective.* Vitasta Publishing Pvt Ltd.

The Right of Women to Menstrual Leave and Free Access to Menstrual Health Products Bill, 276 of 2022, Parliament, Government of India (2022). Introduced in 2017, reintroduced in 2022 https://sansad.in/getFile/BillsTexts/LSBillTexts/Asintroduced/276%20of%202022%20as%20introduced84202375127PM.pdf?source=legislation

Appendices

Appendix A

Interview Guide for Gynaecologists/ Women's Health Doctors

Department of Human Resource Development
Veer Narmad South Gujarat University

ICSSR Project on Women and Menstrual Hygiene in Gujarat
Questions for Gynaecologist and GPs dealing with reproductive health

1 Which type of issues regarding menstruation do you come across?

 i Age – group wise
 ii Marital status wise
 iii Before childbirth, after childbirth

2 Do they discuss frankly or you have to probe?
3 What difficulties do you face in eliciting information? What inhibition?
4 How much do they spend on medical treatment? (Consultancy + medicines)
5 How many go for surgeries? (Cost – pre and post) and which age group?
6 What are the hygienic v/s unhygienic practices according to medical practitioner?
7 What practices (pad, tampons, cloth, or anything else) are frequently observed?
8 Is the menstruation problems linked with malnutrition/anemia/unhygienic practices or any other reasons?
9 Socio economic status of the patients (urban/rural).
10 Anything special.

Appendix B
Questionnaire (English Version)

Department of Human Resource Development

Veer Narmad South Gujarat University, Surat

Schedule for assessing Menstrual Hygiene Practices
(An ICSSR-Sponsored Research Project)

Questionnaire No.: ————— Field Investigator's Initials: —————

Declaration of Ethical Practices   Date: —————

The study examines the practices of menstrual hygiene, beliefs and taboos associated with it. The objective is to get insights into possible direct and indirect costs associated with menstrual hygiene, which might influence the decisions pertaining to practices of menstrual hygiene. Confidentiality and anonymity in publication of findings and the data will be used for academic purpose only.

———————————————

(Signature of the field investigator)

Explicit consent of the Respondent

I have no objection in giving my responses to these questions.

———————————————

(Signature/name of the respondent)

Name of the Respondent: ————————————————————

Contact Information

Village/City/Town: ——————— Taluka: ———————

District: ——————— If urban, address ———————

Phone number: _______________________

Please write appropriate codes/actual values for the responses in the space provided.

Socioeconomic Profile and Habits

Q. No.	Questions (please write the appropriate code/exact value)	Response
1	**Age** (in completed years)	
2	**Occupation** School going student [1] College going student [2] Manual Labour – Outdoors [3] Other work – Outdoors [4] Working – Indoors [5] Non-Working [6]	
3	**Education** Not literate [1] Literate (with or without schooling) [2] Up to Primary [3] Up to Secondary [4] Up to 12th Grade [5] Higher [6]	
4	**Marital Status** Ever married [1] Currently married [2] Never married [3]	
5	If ever/currently married, then age at first marriage	
6	If ever/currently married, number of children	
7	**Family type** Nuclear [1] Joint [2] Extended [3]	
9	**Number of Family Members**	
9a	Males	
9b	Females	
10	**Region** Rural [1] Urban [2]	

Q. No.	Questions (please write the appropriate code/exact value)	Response
11	**Monthly income (Rs.)**	
12	**House Type** Flat [1]　　　　Apartment [2]　　　　Row house [3] Bungalow [4]　　Kachcha House [5]	
13	**No. of Rooms in the house (excluding kitchen)**	
14	**Does the house have a separate kitchen?** Yes [1]　　　　No [0]	
15	**Access to toilet facilities** In house/　　　Community toilet [2]　　　No facility [3] 　IHHL [1]	
16	**Category** General [1]　　SC [2]　　　ST [3]　　　OBC [4]	
17	**Religion** Hindu [1]　　Muslim [2]　　Christian [3]　　Sikh [4] Parsi [5]　　Jain [6]　　Other, Please Specify [7] _______________	
18	**Food Habits** Vegetarian [1]　　Eggetarian [2]　　Non-Vegetarian [3] Vegan [4]	
19	**Addiction**	
19a	**Self** No addiction [1]　　Alcohol [2] Smoking/Vaping [3]　　Chewing tobacco [4]　　Drugs [5]	
19b	**Family Members** No addiction [1]　　Alcohol [2] Smoking/Vaping [3]　　Chewing tobacco [4]　　Drugs [5]	

Awareness, Believes, and Taboos about Menstrual Cycles

Awareness about Menstrual Cycles

Q. No.	Questions (please write the appropriate code/exact value)	Response
20	**How did you first come to know about menstrual cycles?** Mother [1] Sister [2] Friends [3] Teachers [4] Relative [5] Media [6] Internet [7] Any other source, please specify [8] _______________________	
21	**Who guided you during your first menstrual cycle?** Mother [1] Sister [2] Friends [3] Teachers [4] Relative [5] Media [6] Internet [7] Someone else, please specify [8] _______________________	
22	**Who guided you during your initial menstrual cycles?** Mother [1] Sister [2] Friends [3] Teachers [4] Relative [5] Media [6] Internet [7] Someone else, please specify [8] _______________________	
23	**Age at menarche (first period)**	
24	**Are you aware that following changes take place at puberty**	
24a	Physical changes: Breast development Yes [1] No [0]	
24b	Physical changes: Pubic hair growth Yes [1] No [0]	
24c	Physical changes: Hair growth in the armpits Yes [1]] No [0]	
24d	Overall physical changes in the body Yes [1] No [0]	
24e	Hormonal Changes Yes [1] No [0]	
24f	Mood changes Yes [1] No [0]	
24g	Sign of readiness of the body for reproduction Yes [1]] No [0]	
24h	Realisation of some medical conditions, if any Yes [1] No [0]	
24i	Any other, please specify [1] _______________________ Don't know [2]	
25	**How many days between two cycles is normal?**	
26	**Source of menstrual blood** Reproductive tract/Vagina [1] Urinary tract [2] Womb [3] Stomach [4] Don't know [5]	

Beliefs about Menstrual Cycles

Strongly believe [1]   Believe [2]   Haven't thought about [3]
Don't believe [4]   Strongly disbelieve [5]

Q. No.	Belief	Response
27	Menstruation is a disease.	
28	A girl/woman becomes impious during her menstrual cycle.	
29	Absence of menstrual cycles means pregnancy.	
30	Menstrual blood contains dangerous substances.	
31	Pain during menstruation indicates sickness.	
32	Irregularity in cycles is a disease.	
33	Regularity in cycles with frequency of less than a month is a disease.	
34	Regularity in cycles with frequency of more than a month is a disease.	
35	Any other, Please specify _________________________	

Taboos about Menstrual Cycles

Cannot do [1]   Can do but I don't [2]   Can do and I do [3]

Q. No.	Taboos	Response
36	Attend religious ceremonies	
37	Perform religious ceremonies	
38	Offer prayers	
39	Go to temple/church/mosque	
40	Do household work	
41	Cook food	
42	Touching stored food	
43	Live the routine, don't observe seclusion	
44	Sleeping on usual bed	
45	Play outdoors/go to work	
46	Dance	
47	Work involving physical strain	
48	Any other, Please specify _________________________	

Hygiene Practices during Menstrual Cycle

Q. No.	Practices	Response
48	**Which of the menstrual protection products have you heard of?**	
48a	Cloth Yes [1] No [0]	
48b	Cotton Pads Yes [1] No [0]	
48c	Disposable sanitary napkin Yes [1] No [0]	
48d	Tampon Yes [1] No [0]	
48e	Menstrual Cup Yes [1] No [0]	
49	**What do you normally use during your menstrual cycle?** Cloth from saree/old clothes [1] Cotton pads [2] Disposable sanitary napkins [3] Tampons [4] Menstrual Cups [5] Natural Materials (Mud, Cow Dung, Leaves) [6] Any other, please specify [7] _____________________ Nothing [8] Criterion for product selection	
50	If the response to Q.49 is [9], probe. How do you manage *and then go to Q.76?* Otherwise continue.	
51	Do you use more than one product during a menstrual cycle? Yes [1] No [0] If no, skip Q.52, otherwise, continue.	
52	Please mention all the products used by you during a menstrual cycle	
52a	Cloth Yes [1] No [0]	
52a	Cotton Pads Yes [1] No [0]	
52b	Disposable sanitary napkin Yes [1] No [0]	

Q. No.	Practices	Response
52c	Tampon Yes [1] No [0]	
52d	Menstrual Cup Yes [1] No [0]	
52e	Natural Materials (Mud, Cow Dung, Leaves) Yes [1] No [0]	
53	Are you aware of the Government Initiative of sanitary napkins at subsidised rates? Yes [1] No [0]	
54	Where do you usually/normally buy the <<normally used product>> from? Shop in local area [1] Departmental store/superstore [2] Anganwadi [3] ASHA Worker [4] Women centric NGO [5] Any other place, please specify [6] _______________	
55	Who normally buys the <<normally used product>> for you? Self [1] Mother [2] Grandmother [3] Aunt [4] Sister [5] Friend [6] Mother-in-law [7] Sister-in-law [8] Husband [9] Any other person, please specify [10] _______________ If the answer to this question is anything other than [1], go to Q.58a, otherwise continue.	
56	Do you hesitate to ask for the <<normally used product>> from the seller? Yes [1] No [0]	
57	Does anyone accompany you for buying the <<normally used product>>? Yes [1] No [0] If the answer to this question is [0], then go to Q.58, otherwise continue.	
58	Who normally accompanies you to buy the <<normally used product>>? Mother [1] Grandmother [2] Aunt [3] Sister [4] Friend [5] Mother-in-law [6] Sister-in-law [7] Husband [8] Any other person, please specify [9] _______________	
59a	You are using <<product name1>> since (months/years)	
59b	You are using <<product name2>> since (months/years)	
60	Have you changed brands for <<product name1>>? Yes [1] No [0] If response to this question is [0], skip Q.61, otherwise continue.	
61	Reasons for change of brands <<product 1>>	
61a	Improved quality Yes [1] No [0]	

Q. No.	Practices	Response
61b	Enhanced durability Yes [1] No [0]	
61c	Increased comfort Yes [1] No [0]	
61d	Rash-proof Yes [1] No [0]	
61e	Better fragrance Yes [1] No [0]	
61f	More pocket friendly Yes [1] No [0]	
61g	Environmentally friendly Yes [1] No [0]	
61h	Any other, please specify ___________________ Yes [1] No [0]	
62	Have you changed brands <<product name2>> Yes [1] No [0] If response to this question is [0], skip 63, otherwise continue.	
63	Reasons for change of brands <<product name 2>>	
63a	Improved quality Yes [1] No [0]	
63b	Enhanced durability Yes [1] No [0]	
63c	Increased comfort Yes [1] No [0]	
63d	Rash-proof Yes [1] No [0]	
63e	Leak Proof Yes [1] No [0]	
63f	More pocket friendly Yes [1] No [0]	
63g	Environmentally friendly Yes [1] No [0]	
63h	Any other, please specify ___________________ Yes [1] No [0]	
64	Sanitary napkins/tampons users' methods of disposal	
64a	Burn it Yes [1] No [0]	
64b	Throw it in Routine waste Yes [1] No [0]	

Q. No.	Practices	Response
64c	Flush Yes [1]　No [0]	
64d	Incinerator at school/college/work place Yes [1]　No [0]	
64e	Any other, please specify ________________ Yes [1]　No [0]	
Q. 65 to 67c is for the Menstrual Cup Users		
65	When do you plan to change your present menstrual cup? (after ________ months/years?	
66	Have you ever disposed a menstrual cup? Yes [1]　No [0]	
67	How have you disposed/plan to dispose your menstrual cup?	
67a	Throw it in Routine waste Yes [1]　No [0]	
67b	Dispose it at facilities for recycling Yes [1]　No [0]	
67c	Any other, please specify: ________________ Yes [1]　No [0]	
68	Is there any Transition in products used? Yes [1]　No [0] If the response to this question is [0], skip questions 69 and 70, otherwise, continue.	
69a	**Transition from** Cloth from saree/old clothes [1]　Locally Prepared Napkin [2] Disposable sanitary napkins [3]　Menstrual cups [4] Tampons [5]　Cotton pads [6] Natural Materials (Mud, Cow Dung, Leaves) [7] Any other, please specify [8] ________________ Nothing [9]	
69b	**Transition to** Cloth from saree/old clothes [1]　Locally Prepared Napkin [2] Disposable sanitary napkins [3]　Menstrual cups [4] Tampons [5]　Cotton pads [6] Natural Materials (Mud, Cow Dung, Leaves) [7] Any other, please specify [8] ________________	
70	Reasons for change in products	
70a	Improved hygiene Yes [1]　No [0]	

Q. No.	Practices	Response
70b	Enhanced durability Yes [1] No [0]	
70c	Increased comfort Yes [1] No [0]	
70d	Rash-proof Yes [1] No [0]	
70e	Keeps the foul smell away Yes [1] No [0]	
70f	More pocket friendly Yes [1] No [0]	
70g	Environment friendly Yes [1] No [0]	
70h	Any other, please specify: _______________________ Yes [1] No [0]	
71	Do your family members' have a say in what product to use? Yes [1] No [0]	
72	Does your husband (only to be asked to currently married) have a say in what product to use? Yes [1] No [0]	
73	How often do you clean your private parts? Once a day [1] Twice a day [2] Every time you go to the toilet [3] Every time you change your pad/tampon/remove your menstrual cup [4] I don't clean my private parts during menstruation [5] If the response to this question is [5], skip Q.74, otherwise continue.	
74	What do you use to clean your private parts? Only Water [1] Bathing soap [2] Antiseptic/Disinfectant [3] Intimate hygiene products (e.g. VWash etc.) [4] Anything else, please specify _______________ [5]	
75	What do you use to clean your hands after using the washroom? Only Water [1] Mud/Sand [2] Bathing soap [3] Hand wash [5] Antiseptic/Disinfectant/Hand sanitiser [6] Anything else, please specify _______________ [7]	
76	If you are using reusable pads/pieces of cloth, where do you dry after washing them? Outside house in Sunlight [1] Inside the house in a closed room [2] Inside the house beneath a cloth [3] Don't use reusable pads/pieces of cloth [4]	

Cost of Menstrual Hygiene
Product Cost (Direct Costs)

Q. No.	Costs and its Components	Response
Questions 77 to 79e are for sanitary napkin/tampon users		
77	Price per packet	
78	Number of pads/tampons in a packet	
79	Pads/tampons usage per day	
79a	Day1	
79b	Day2	
79c	Day3	
79d	Day4	
79e	Day5	
79f	Day6	
Questions 80 to 82 are for menstrual cup users		
80	Price of the menstrual cup	
81	Is this your first cup Yes [1] No [0] If the response to this question is [1], then skip Q.82, otherwise continue.	
82	How often do you change menstrual cups (months/years)?	
83	What is the cost of the antiseptic/disinfectant/intimate hygiene product used for cleaning your private parts?	
84	How long does it last (months)?	
85	What is the cost of the handwash/antiseptic/disinfectant/hand sanitiser product used for washing your hands after washroom use?	
86	How long does it last (months)?	

Psychological and Physical Problems during Menstrual Cycle and Resultant Costs
(PMS = Pre-Menstrual Syndrome, DM = During Menstruation)
Please mark Yes [1] or No [0]

Q. No.	Psychological Symptoms	PMS	DM
87	Anxiety		
88	Confusion		
89	Forgetfulness		

Q. No.	Psychological Symptoms	PMS	DM
90	Irritability		
91	Mood swings		
92	Anger feats		
93	Difficulty in concentrating		
94	Depression		
95	Anything else, please specify ________________		

Q. No.	Physical Symptoms	PMS	DM
96	Head ache		
97	Stomach ache		
98	Muscle fatigue		
99	General fatigue		
100	Dizziness		
101	Giddiness		
102	Sweating/palpitation		
103	Nausea/Vomiting/Diarrhoea		
104	Fainting		
105	Tenderness in breasts		
106	Bloating		
107	Anything else, please specify ________________		
If response to all of the above questions is [0], then go the next section (Absenteeism), otherwise, continue.			
108	Ever consulted a doctor for any of these symptoms? Yes [1] No [0] If the response to this question is [0], then skip Q.107 to 112, otherwise continue.		
109	Number of times in past one year		
110	Consultation charges per check-up		
111	What was the spell (number of days) of treatment?		
112	How often (number of days) do you consult the doctor in each spell?		
113	Does the doctor prescribe medicines? Yes [1] No [0] If the response to this question is [0], then skip Q.112, otherwise continue.		
114	How much does the medicine cost per spell of treatment?		

Q. No.	Activities	Response
115	Any effect on any of the following essential household work because of psychological or physical symptoms? Please write appropriate option (code) in front of each of the activities. Can't do it/not allowed to do [1] other female family members do it [2] Any other family members do it [3] hire paid help/buy food from outside [4] No effect [5] If the answer to all of the following is [1] to [3], skip Q.116, otherwise continue.	
115a	Cooking	
115b	Cleaning utensils	
115c	Washing clothes	
115d	Fetching water	
116	Expenditure on paid help for each of the following activities.	
116a	Cooking	
116b	Cleaning utensils	
116c	Washing clothes	
116d	Fetching water	

Cost resulting from Poor Menstrual Hygiene Practices

(If the value in the column "number of times so far" is 0, please write the code 99 for Not Applicable in rest of the columns for a given medical condition.)

Q. No.	Medical conditions due to poor menstrual hygiene practices? (Write zero, if not experienced)	Number of times so far	Home Remedies	Doctor's Consultation	Cost of Medicines	Number of days lost
117	Itching in private parts/vaginal tract					
118	Vaginal Infection					
119	Herpes in private parts					
120	Rashes in/around the vaginal tract					

Q. No.	Medical conditions due to poor menstrual hygiene practices? (Write zero, if not experienced)	Number of times so far	Home Remedies	Doctor's Consultation	Cost of Medicines	Number of days lost
121	Internal swelling resulting in pain in pelvic region					
122	Moniliasis (fungal infection)					
123	Abscess					
124	Any Other, Please Specify ______					
125	Number of hospitalisation episodes (if any) If the response to this question is 0, then mention 99 for not applicable in response to Q.126, otherwise write actual costs.					
126	Total expenditure per hospitalisation episode (if any)					

Absenteeism during Menstrual Cycles and Associated Costs

(This section applies to working women and students)
Please mark Yes [1] or No [0]

Q. No.	Questions	Response
127	Reasons for absenteeism during menstrual cycle	
127a	I don't remain absent. If the response to 127a is [1], then go to Q.128	
127b	I am afraid of staining my clothes.	
127c	I am afraid of others making fun of me.	
127d	Periods causes pain.	
127e	Makes me feel uncomfortable/tired.	
127f	No toilet facilities at school/work.	
127g	Absence of clean toilets.	
127h	No places for girls to wash and change at school.	
127i	No place to dispose used absorbents.	
127j	Any other, please specify ______	
128	No of days you remain absent from school/college/work? If the response to 125a is [1], then the response to this question is 0.	
129	No of days you get paid offs If the response to 126 is [0], mention 99 for "Not applicable" in the response to this question.	

130 Are you aware about any medical conditions you had/currently having?

____

____

____

131 Does your husband (only to married women)/other family members support you during your menstrual cycles? How?

____

____

____

132 Are men (if there are any) in your family aware about menstruation/ menstrual hygiene and related issues?

____

____

____

Field Investigator's Remarks: _______________________

____

____

____

____

____

Thank you for sparing your valuable time

Appendix C
An Interview Guide for Girls/Women

Department of Human Resource Development

Veer Narmad South Gujarat University

ICSSR Project on Women and Menstrual Hygiene in Gujarat

In-Depth Interviews for Girls/Women

The basic purpose of undertaking in-depth interviews is getting insights on the problems faced by the women during their span of menstruation (right from menarche) and their coping mechanisms:

Broad Issues to be discussed

- Socioeconomic background (especially, age and marital status).
- Experience narration of the problems faced.
- Coping strategies and the extent to which the problems are addressed.
- Ever consulted a doctor for coping up with problems. Experience narration.
- Any change in problems before and after childbirth (for married women).
- The present state of their menstruation-related problems.

Appendix D
Interview Guide for NGOs

Department of Human Resource Development
Veer Narmad South Gujarat University

ICSSR Project on Women and Menstrual Hygiene in Gujarat
Questions for NGOs

1 What services are you providing for menstrual hygiene?
2 What made you think of undertaking this work?
3 Have you faced any difficulties? If so, what kind of? How did you handle those difficulties?
4 How do you manage financial resources?
5 What age group and region do you cater to?
6 Are there any age-specific problems?
7 Have you observed any change in the belief pattern/practices during menstruation after you started the work? If so, what?
8 Socio economic status of your beneficiaries (urban/rural).
9 Anything special.

Appendix E

Respondents by Districts and Region

Table E.1 Number of Respondents by District and Region

District	Rural	Urban	Total
Ahmedabad	3	18	21
Anand	104	41	145
Aravalli	1	0	1
Banaskantha	1	0	1
Bharuch	11	12	23
Bhavnagar	0	4	4
Chhota Udepur	1	0	1
Dahod	4	0	4
Gandhinagar	0	1	1
Godhra	0	3	3
Junagadh	1	0	1
Kachchh	2	1	3
Kheda	50	51	101
Mehsana	2	0	2
Narmada	4	2	6
Navsari	26	35	61
Patan	1	0	1
Rajkot	0	1	1
Surat	151	293	444
Surendranagar	1	0	1
Tapi	93	5	98
Vadodara	2	73	75
Valsad	5	8	13
<NA>	9	5	14

Appendix F
Technical Notes on Model Selection using Stepwise Regression and Outputs

When there are many predictors, and different models using different combination of predictors are possible, Akaike Information Criteria (AIC) is one way of selecting the model with maximum information; the other being Bayesian Information Criteria (BIC). Apart from information criteria, there are significance criteria, penalised likelihood and change-in-estimate criteria for model selection (Heinze et al., 2018). The estimations are undertaken in R software. By default, R uses AIC for model selection in stepwise regression. We have made use of forward stepwise regression. Other stepwise regressions are backward stepwise and mixed stepwise regression approaches.

$$AIC = -2\log L(x\,|\,\hat{\beta}) + 2k,$$

where

$\log L$ = Log Likelihood
x = set of predictors,
$\hat{\beta}$ = set of estimated parameters, and
k = number of predictors.

AIC could, therefore, be positive or negative and the lowest AIC score indicates the model with maximum information.

The summarised results for model comparison using AIC, for awareness, beliefs and taboos are given in Tables F.1–F.3.

Table F.1 Model Selection Summary for Awareness Scores Model using AIC Criteria

Steps	Regression	AIC
Start	Awareness ~ 1	1956.35
Step 1	Awareness ~ Occupation	1801.31
Step 2	Awareness ~ Occupation + Region	1728.98
Step 3	Awareness ~ Occupation + Region + Development.Status	1708.59
Step 4	Awareness ~ Occupation + Region + Development.Status + Family.type	1698.58
Step 5	Awareness ~ Occupation + Region + Development.Status + Family.type + Education	1691.45
Step 6	Awareness ~ Occupation + Region + Development.Status + Family.type + Education + Marital.Status	1691.34

Table F.2 Model Selection Summary for Belief Scores Model using AIC Criteria

Steps	Regression	AIC
Start	belief.score ~ 1	−3248.45
Step 1	belief.score ~ Education	−3304.64
Step 2	belief.score ~ Education + Occupation	−3348.76
Step 3	belief.score ~ Education + Occupation + Region	−3362.7
Step 4	belief.score ~ Education + Occupation + Region + Development.Status	−3374.47

Table F.3 Model Selection Summary for Taboo Scores Model using AIC Criteria

Steps	Regression	AIC
Start	taboo.score ~ 1	−3024.79
Step 1	taboo.score ~ Religion	−3096.79
Step 2	taboo.score ~ Religion + Region	−3165.54
Step 3	taboo.score ~ Religion + Region + Education	−3200.98
Step 4	taboo.score ~ Religion + Region + Education + Occupation	−3226.62
Step 5	taboo.score ~ Religion + Region + Education + Occupation + Family.type	−3233.95
Step 6	taboo.score ~ Religion + Region + Education + Occupation + Family.type + Development.Status	−3236.44

Reference

Heinze, G., Wallisch, C., & Dunkler, D. (2018). Variable selection—A review and recommendations for the practicing statistician. *Biometrical Journal. Biometrische Zeitschrift, 60*(3), 431–449. https://doi.org/10.1002/bimj.201700067

Appendix G

Formulae for Estimating the Costs Associated with Menstrual Health

Explicit Costs

Explicit cost of using disposable sanitary napkins/tampons

$$C_A = \left(\sum_{D_i=1}^{6} N_A \right) \left(\frac{P_P}{N_P} \right) \times 12 \qquad \text{(G.1)}$$

C_A is the yearly cost of absorbent (disposable sanitary napkins/tampons),
D_i is the day of the menstrual cycle,
N_A is the number of absorbents in a packet,
P_P is the price per packet, and
N_P is the number of sanitary napkins/tampons per packet

Explicit cost of Allied Products

$$C_{AP} = \left(\left(\frac{P_{AP}}{D_{AP}} \right) \Big/ N_{AP} \right) \times 12 \qquad \text{(G.2)}$$

C_{AP} is the yearly cost of allied products.
P_{AP} is the price of allied product,
D_{AP} is the duration (in months) the product lasts, and
N_{AP} is the number of persons with whom the product is shared

Indirect and Implicit Costs

Costs of consultation charges per year

$$C_c = N_c x N_s x P_c \qquad \text{(G.3)}$$

where,
C_c is the yearly cost of consultation,
N_c is the number of consultations per spell,
N_s is the number of spells per year, and
P_c is the per unit price of consultation (doctor's fees per each consultation)

Medicines costs per year

$$C_m = N_s x P_m \qquad \text{(G.4)}$$

where,
C_m is the yearly cost of medicines,
N_s is the number of spells of seeking medical advice per year, and
P_m is the average per spell expenditure on medicines.

Index

For Product Safety Concerns and Information please contact our EU
representative GPSR@taylorandfrancis.com
Taylor & Francis Verlag GmbH, Kaufingerstraße 24, 80331 München, Germany